SpringerBriefs in Applied Sciences and Technology

Manufacturing and Surface Engineering

Series Editor

Joao Paulo Davim, Department of Mechanical Engineering, University of Aveiro, Aveiro, Portugal

This series fosters information exchange and discussion on all aspects of manufacturing and surface engineering for modern industry. This series focuses on manufacturing with emphasis in machining and forming technologies, including traditional machining (turning, milling, drilling, etc.), non-traditional machining (EDM, USM, LAM, etc.), abrasive machining, hard part machining, high speed machining, high efficiency machining, micromachining, internet-based machining, metal casting, joining, powder metallurgy, extrusion, forging, rolling, drawing, sheet metal forming, microforming, hydroforming, thermoforming, incremental forming, plastics/composites processing, ceramic processing, hybrid processes (thermal, plasma, chemical and electrical energy assisted methods), etc. The manufacturability of all materials will be considered, including metals, polymers, ceramics, composites, biomaterials, nanomaterials, etc. The series covers the full range of surface engineering aspects such as surface metrology, surface integrity, contact mechanics, friction and wear, lubrication and lubricants, coatings an surface treatments, multiscale tribology including biomedical systems and manufacturing processes. Moreover, the series covers the computational methods and optimization techniques applied in manufacturing and surface engineering. Contributions to this book series are welcome on all subjects of manufacturing and surface engineering. Especially welcome are books that pioneer new research directions, raise new questions and new possibilities, or examine old problems from a new angle. To submit a proposal or request further information, please contact Dr. Mayra Castro, Publishing Editor Applied Sciences, via mayra.castro@springer.com or Professor J. Paulo Davim, Book Series Editor, via pdavim@ua.pt

More information about this series at http://www.springer.com/series/10623

Kaushik Kumar · Divya Zindani ·
J. Paulo Davim

Industry 4.0

Developments towards the Fourth Industrial
Revolution

Springer

Kaushik Kumar (iD)
Department of Mechanical Engineering
Birla Institute of Technology
Mesra, Ranchi, Jharkhand, India

Divya Zindani
Department of Mechanical Engineering
National Institute of Technology Silchar
Silchar, Cachar, Assam, India

J. Paulo Davim
Department of Mechanical Engineering
University of Aveiro
Aveiro, Portugal

ISSN 2191-530X   ISSN 2191-5318 (electronic)
SpringerBriefs in Applied Sciences and Technology
ISSN 2365-8223   ISSN 2365-8231 (electronic)
Manufacturing and Surface Engineering
ISBN 978-981-13-8164-5   ISBN 978-981-13-8165-2 (eBook)
https://doi.org/10.1007/978-981-13-8165-2

This Springer imprint is published by the registered company Springer Nature Singapore Pte Ltd.
The registered company address is: 152 Beach Road, #21-01/04 Gateway East, Singapore 189721, Singapore

Preface

The authors are pleased to present the book *Industry 4.0—Developments towards the Fourth Industrial Revolution* under the book series *Manufacturing and Surface Engineering*. The book title was chosen understanding the current importance of Industry 4.0 as well as in future for the industrial and manufacturing world.

After the Renaissance, the Industrial Revolution with technology-driven outlook was the giant step for global development and prosperity. The Industrial Revolution started in around 1750 with I 1.0 (First Industrial Revolution) where mechanical production facilities, e.g. steam engine, spinning wheel, water wheel, were introduced for better and more productivity. A century later in around 1850, manufacturing with the help of electricity, assembly lines, conveyor belts, etc., introduced the concept of mass production and was designated as Second Industrial Revolution (I 2.0). Third Industrial Revolution (I 3.0), in the nineteenth century, saw the integration of manufacturing with electronics and provided the era of NC, CNC, DNC classes of automated production machinery.

Presently with globalization and open market economy, the market has become consumer driven or customer dictated. This has given rise to the 4th Industrial Revolution or I 4.0. This has initiated amalgamation of Internet, information and communication technologies (ICTs) and physical machinery with the coinage of words like Internet of things (IoT), industrial Internet of things (IIoT), cobot (collaborative robot), big data, cloud computing, virtual manufacturing, 3D printing finding their way into our daily life. I 4.0 has been designed towards the development of a new generation of smart factories or currently coined as 'customized or tabletop factories' with increased production flexibility allowing personalized and customized production of articles in a lot as small as a single unique item. Hence, today facilities are to be provided to a customer situated at one side of the globe to control and monitor his product being produced in a manufacturing unit available at another side of the world.

This book is primarily intended to provide researchers and students with an overview of the buzzword 'Industry 4.0' that promises to increased flexibility in manufacturing in tandem with the mass communication, improved productivity and better quality. The book provides an overview of topics associated with Industry 4.0, i.e. intelligent manufacturing, process planning, assessment of product development opportunities, tools, aspects of risk management, education and qualification requirements, socio-technical considerations and sustainability of business models in I 4.0 era.

This book contains six chapters. Chapter 1 (Intelligent Manufacturing) introduces the concept of Industry 4.0, i.e. the next generation of the industry to the readers. It attempts to illuminate the readers with the associated key concepts such as intelligent manufacturing, cloud manufacturing and Industry 4.0 and key enabling technologies such as big data analytics, cyber-physical systems, Internet of things, information and communication technology and cloud computing. The chapter also provides an overview on platforms in intelligent manufacturing and on few emerging trends in intelligent manufacturing. This chapter finally concludes with future perspectives for intelligent manufacturing in the context of Industry 4.0.

The next chapter, Chap. 2 (Process Planning in Era 4.0), deals with the working environment in Fourth Industrial Revolution. With the constantly changing manufacturing environment, the personnel involved will have to learn the new skills and adapt to these changes that will ultimately aid in the enhancement of their performances. And subsequently, it will result in enhanced productivity, quality of product, reduced manufacturing time and would affect product prices. With the advancements in the manufacturing environment, the concept of mass customization will be easy to realize. The present chapter reflects the changing role of process planner to product planner. Product planner, in the realm of Industry 4.0, is actually software that is connected to other parts in a supply chain. The software is capable of generating order of scheduling, order of operations and process plan using advanced optimization algorithms. This chapter aims to provide an overview of the product planner that aids in automatically planning, scheduling and operation sequencing.

Chapter 3 (Requirements of Education and Qualification) presents knowledge needs of employees. Industry 4.0 (I 4.0) has resulted in enhanced productivity and mass customization through new technological advances and methodologies. The new approaches need to be flawlessly introduced in the company given the resistance from the workforce to accept the same. The transition, however, cannot be accepted overnight. The main reasons being high capital investments and lack of required skills within the workforce. The very aim of the present chapter is to identify the various job roles in the companies ready to accept I 4.0 environment.

The management of production and manufacturing processes has been automatized in the Industry 4.0 era. This has been possible with the development of more complex IT infrastructure resulting in modified network and frameworks. However, with the modifications comes the risk. The framework comprises connections between manufacturing systems, objects, activities and humans. Owing to the connections and real-time processing of data, frameworks have become more complex. Furthermore, from time to time, there is always a requirement to upgrade the existing management infrastructure, production infrastructure and the related technologies. This is because of the volume of data generated and the demand for mass customization of products. As such the risks associated under such environment is inevitable. The theme of the present chapter, Chap. 4 (Risk Management Implementation), is to illuminate its readers with the aspects of risk management in Industry 4.0 era and its implementation through a dedicated framework.

All the above actions would surely alter the socio-technical conditions, and hence, Chap. 5 (Socio-technical Considerations) reviews on the cost and the efficiency gains associated with the infrastructure and therefore the implementation of Industry 4.0. The feasibility of its adoption by the manufacturing units is argued in the perspective of the internal capabilities of Industry 4.0 to be the driving agent for the creation of the competitive advantage. The importance of lean manufacturing as a key supporter for Industry 4.0 implementation has been presented in this chapter. Lean methods are the key enablers for the successful implementation of I 4.0. The chapter also puts forth to discuss the technical aspects of Industry 4.0 and the importance of the socio-technical requirements resulting in the successful implementation of Industry 4.0.

Sustainable business models have been established in the digital and automated environment. However, the existence of such business models is still not mainstream. Sustainable manufacturing has ample opportunities in the market as there is a need to design products that promotes the concept of repair, recycling and longevity. The efficiency of the models is not only the sole objective, but these models also take into consideration the lesser use of raw materials and the recycling of manufactured products. All these proponents of the sustainable business model result in changing of the value proposition and customer relationship and also affects the supply chain. The last chapter of the book, Chap. 6 (Sustainable Business Scenarios in 4.0 Era), illuminates the readers with the potential scenarios for such business models in the backdrop of Industry 4.0.

First and foremost we would like to thank God for everything. It was His blessings that this work could be completed to our satisfaction. You have given the power to believe in passion, hard work and pursue dreams. This could never have been done without the faith in You, the Almighty.

We would like to thank our grandparents, parents and relatives for allowing us to follow our ambitions. Our families showed patience and tolerated us for taking yet another challenge which decreases the amount of time we could spend with them. They were our inspiration and motivation. Our efforts will come to a level of satisfaction if the professionals concerned with all the fields related to Industry 4.0 get benefitted.

We would also like to thank the reviewers, the editorial board members, project development editor and the complete team at Springer Nature. Throughout the process of writing this book, many individuals, from different walks of life, have taken time out to help. Last but not least, we would like to thank them all for encouraging us. The project would have got shelved without their support.

Mesra, Ranchi, Jharkhand, India   Kaushik Kumar
Silchar, Cachar, Assam, India   Divya Zindani
Aveiro, Portugal   J. Paulo Davim

Contents

About the Authors

Kaushik Kumar B.Tech (Mechanical Engineering, REC (Now NIT), Warangal), MBA (Marketing, IGNOU) and Ph.D (Engineering, Jadavpur University), is presently an Associate Professor in the Department of Mechanical Engineering, Birla Institute of Technology, Mesra, Ranchi, India. He has 16 years of Teaching & Research and over 11 years of industrial experience in a manufacturing unit of Global repute. His areas of teaching and research interest are Conventional and Non-conventional Quality Management Systems, Optimization, Non-conventional machining, CAD/CAM, Rapid Prototyping and Composites. He has 9 Patents, 26 Book, 13 Edited Book, 42 Book Chapters, 136 international Journal publications, 21 International and 8 National Conference publications to his credit. He is on the editorial board and review panel of 7 International and 1 National Journals of repute. He has been felicitated with many awards and honours.

Divya Zindani (BE, Mechanical Engineering, Rajasthan Technical University, Kota), M.E. (Design of Mechanical Equipment, BIT Mesra), presently pursuing PhD (National Institute of Technology, Silchar). He has over 2 years of Industrial experience. His areas of interests are Optimization, Product and Process Design, CAD/CAM/CAE, Rapid prototyping and Material Selection. He has 1 Patent, 4 Books, 6 Edited Books, 18 Book Chapters, 2 SCI journal, 7 Scopus Indexed international journal and 4 International Conference publications to his credit.

J. Paulo Davim received his Ph.D. in mechanical engineering in 1997, M.Sc. in mechanical engineering (materials and manufacturing processes) in 1991, Licentiate degree (5 years) in mechanical engineering in 1986 from the University of Porto (FEUP), the Aggregate title from the University of Coimbra in 2005 and D.Sc. from London Metropolitan University in 2013. He is EUR ING by FEANI and senior chartered engineer by the Portuguese Institution of Engineers with MBA and specialist title in engineering and industrial management. Currently, he is professor at the Department of Mechanical Engineering of the University of Aveiro. He has more than 30 years of teaching and research experience in manufacturing, materials and mechanical engineering with special emphasis in machining and tribology. Recently, he has also interest in management/industrial engineering and higher education for sustainability/engineering education. He has received several scientific awards. He has worked as evaluator of projects for international research agencies as well as examiner of Ph.D. thesis for many universities. He is the editor in chief of several international journals, guest editor of journals, editor of books, series editor of books and scientific advisor for many international journals and conferences. At present, he is an editorial board member of 25 international journals and acts as reviewer for more than 80 prestigious Web of Science journals. In addition, he has also published as editor (and co-editor) for more than 100 books and as author (and co-author) for more than 10 books, 80 book chapters and 400 articles in journals and conferences (more than 200 articles in journals indexed in Web of Science/h-index 45+ and SCOPUS/h-index 52+).

Chapter 1
Intelligent Manufacturing

1.1 Introduction

Industry 4.0 aims to create factories for manufacturing, wherein the manufacturing environment is made intelligent through the employability of cloud computing, Internet of Things and cyber-physical systems (Lee et al. 2015). The different manufacturing systems are intelligent such that they are able to transform the physical world into digital twin or cyber twin and therefore easily monitor and take decisions effectively of the associated processes. The cooperation with machines, sensors and humans as well as the real-time communication between these makes the task a tad easier (Wang et al. 2016a, b). Industry 4.0, therefore, aims to bring about transformation in industry value chains, associated models of business and value chains of production by combining embedded production system with intelligent systems of production.

Manufacturing systems are upgraded to being intelligent in the gamut of Industry 4.0. The demands of the global and dynamic market are now met comfortably with the proposition of intelligent manufacturing that takes advantage of advanced manufacturing and information technologies in order to achieve manufacturing processes that are reconfigurable, smart and flexible (Shen and Norrie 1999). Intelligent manufacturing environment results in the availability of physical processes and the related information as and when required for the various industries, enterprises and supply chains (Wan et al. 2017; Wang et al. 2016a, b). Some of the key enabling technologies are required to aid machines and devices to vary their behaviour in accordance with the desired requirements, situations or as per the past experiences (McFarlaneb et al. 2003). The underpinning technologies allow for solving the manufacturing problems dynamically and hence the decisions to be made accordingly within the stipulated time frame. This is achieved through the underpinning technologies that aid the direct and real-time communication with the different manufacturing systems. As for instance, the artificial intelligence (AI) technology aids the manufacturing components to learn from past experiences so that intelligent and universally accepted industrial practices could be realized.

K. Kumar et al., *Industry 4.0*, Manufacturing and Surface Engineering,
https://doi.org/10.1007/978-981-13-8165-2_1

Internet of Things enabled manufacturing as well as cloud manufacturing are few other conceptual frameworks for intelligent manufacturing. The present chapter is, therefore, aimed to illuminate and comprehend its readers to the conceptual framework of intelligent manufacturing giving due consideration to the gamut of Industry 4.0. Key enabling technologies such as big data analytics, cyber-physical systems, Internet of Things, information and communication technology and cloud computing have been discussed in this chapter. The chapter also provides an overview of platforms in intelligent manufacturing and on few emerging trends in intelligent manufacturing. The chapter finally concludes with future perspectives for intelligent manufacturing in the context of Industry 4.0.

1.2   Few Concepts

Manufacturing enterprises as well industries lays the foundation of the nation's economy. The living style of the people is significantly influenced by the goods and services produced by the manufacturing enterprises. The key enabling technologies can result in game-changing impacts on the models of manufacturing, related concepts and approaches, business models and logistics or supply chain network. Three most advanced manufacturing technologies, i.e. cloud manufacturing, intelligent manufacturing and IoT-enabled manufacturing have been discussed in this section.

1.2.1   Cloud Manufacturing

Cloud manufacturing is one of the models of advanced manufacturing that aims to comprehensibly share and circulate the manufacturing resources through the support of integrated framework comprising of service-oriented technologies, virtualization, Internet of things and cloud computing (Li et al. 2010; Xu 2012). In the cloud manufacturing milieu, required resources and processes are managed intelligently throughout the product life cycle beginning from its design, manufacturing and maintenance. Therefore cloud computing is often regarded as a parallel and networked intelligent manufacturing system. Cloud manufacturing, therefore, allows for on-demand manufacturing services through the manufacturing cloud as and when desired by the client (Zhang et al. 2014).

Various manufacturing resources and processes are interconnected and connected into the cloud. The resources can be managed automatically and controlled using the various Internet of things technologies such as barcodes that makes it easier to be shared digitally. The conceptual framework of cloud manufacturing is supported by such underpinning technologies that are characteristic of being service oriented and cloud manufacturing. As a result of the supportive and integrated framework, accessing, invoking and implementation of various manufacturing services becomes realistic to the client side as these resources can be encapsulated, virtualized and circulated as and when desired (Wu et al. 2013). Further, the various resources and capacities can be categorized and aggregated through the predefined specific rules. Many different types of manufacturing clouds have been developed to handle the activities pertaining to manufacturing hassle free (Chen et al. 2017). Virtual manufacturing platform has aided the customers to search and invoke the desired services as and when required.

Some of the important major concerns in the realm of cloud manufacturing include modes for cloud deployment, modelling of manufacturing resources and matching between requirements and services. Universal access to the clients can be made through the establishment of virtual manufacturing environment that can aid in the deployment of various types of clouds such as private, public, hybrid and community clouds. In the case of hybrid clouds, multiple deployment modes are offered as the environment is a mix of different kinds of clouds. Therefore, advantages such as flexibility in deployment and cross-business are offered by hybrid clouds (Tao et al. 2014a, b). Assembly lines and other related machining resources are modelled into unified services that can be circulated and shared. An effort in this direction has been made by German Electrical and Electronic Manufacturers' association that have developed Reference Architectural Model Industry (RAMI) model, which references to Industry 4.0 taking care of the management and administration aspects of several devices. The developed model has allowed for consistent usage of data and resources. However, the development of such an environment is challenging as it entails the participation of various types of objects and resources that have heterogeneous characteristics that result in unexpected complexity pertaining to modelling (Wang and Xu 2013). It is quintessential to match various requirements and services within the working environment of cloud manufacturing. Planning, execution and scheduling of services, as well as optimal solution for both the customer and service providers, are encompassed within the matching process (Liu et al. 2017).

1.2.2 Internet of Things (IoT) Enabled Manufacturing

The manufacturing objects in Internet of Things encompassed manufacturing are transformed into being smart manufacturing objects. These smart objects are able to sense, interact and interconnect with one another and therefore carry out functions of manufacturing automatically and adaptively (Zhong et al. 2013a, b). Perceptions in the environment of Internet of things enabled manufacturing for various interactions such as that between humans, machines and humans and machines are realized to be as intelligent (Tao et al. 2014a, b). The application of Internet of Things technologies results in on-demand usage of resources and efficient circulation of objects amongst the stakeholders. Adoption of certain technologies such as the infrastructure for sharing and acquisition of data within the Internet of things enabled manufacturing has further enhanced the manufacturing system performance.

Real-time collection of data as well as sharing of the same amongst the various resources such as jobs, materials, jobs and machines is one of the characteristic feature of Internet of things enabled manufacturing environment (Bi et al. 2014). Some of the underpinning technologies that aids in real-time collection of data as well as its sharing are wireless communication and radio-frequency identification (RFID). The data related to the movement of materials, its traceability and visibility is integrated flawlessly by employing RFID technology (Lu et al. 2006; Zhong et al. 2013a, b). Important manufacturing sites such as warehouses, assembly lines and shop floors are the locations, where the creation of smart objects is accomplished. This is done by deploying the RFID tags to the manufacturing objects at the aforementioned manufacturing locations. Any of the disturbances taking place at these manufacturing sites are detected on real-time basis and are fed back to the manufacturing resources in real time (Huang et al. 2008). Thus, RFID tags aids in enhanced effectiveness for decision-making processes related to the manufacturing and production aspects.

Internet of Things enabled manufacturing have been employed in many real-life cases. As, for instance, a real-time system of production management for motorcycle assembly line was reported and investigated (Liu et al. 2012) and it was revealed that the flexibility of manufacturing was increased. The production system was implemented for Loncin Motor Co., Ltd. and the system aided in the collection of data in real time from the resources such as raw materials and staff. This system resulted in improved trackability, traceability and visibility for items of great importance and interest. Another example is that from Huaiji Dengyung Auto-Parts (Holdings) Co., Ltd. that manufacturers engine valves. This SME adopted the RFID-based solution for shop-floor manufacturing. Efforts were made to integrate the systems of enterprise resource planning and manufacturing execution and therefore to extend the existing RFID-enabled manufacturing system. Another instance wherein the implementation of RFID-enabled manufacturing system was made was for Guangdong Chigo Air Conditioning Co., Ltd (Qu et al. 2012). The RFID-enabled system was employed by the enterprise for material management. Accurate and automatic data related to the manufacturing object was obtained through the RFID-enabled system that aided in real-time object visibility and traceability. Some of the other examples are from product life cycle management enterprises, automotive manufacturers, mould and die making industries, aerospace sector and other related manufacturing partners (Dai et al. 2012).

1.2.3 Intelligent Manufacturing

Intelligent manufacturing encompasses in itself the broader concept of manufacturing that aims to optimize production activities by employing advanced information and technologies of manufacturing (Kusiak 1990). Intelligent science and technology forms the basis of the conceptual framework of intelligent manufacturing that aids in upgradation and management of production and design-related aspects of a typical product. Advanced materials, decision-making models with adaptability, smart sensors, data analytics and intelligent devices facilitate the entire product life cycle of a product (Li et al. 2017). Efficiency as well as quality of production is enhanced which ultimately results in enhanced service level (Davis et al. 2012). Industries are equipped to meet the challenges of dynamic and fluctuating markets which enhances their competitiveness.

The aforementioned conceptual framework can be realized through the conceptual framework of intelligent manufacturing. Such a manufacturing environment is realized through the adoption of new models, forms and technologies that ultimately aids in the transformation of traditional objects to being smart in characteristics. The environment of intelligent manufacturing in the context of Industry 4.0 makes use of service-oriented architecture through the means of Internet. Thus, the framework aids in providing its end users with services that are reconfigurable, flexible, customizable and collaborative. Such characteristics result in an integrated framework, which is so referred to as human–machine manufacturing system (Feeney et al. 2015). With the aid of a highly integrated system, it becomes possible to establish a system in which the technical, managerial as well as organizational levels of a manufacturing enterprise can be combined seamlessly. Festo Didactic cyber-physical factory is one of the examples that employs the intelligent manufacturing system. Within this framework, the factory offers to train vendors, schools and universities in order to facilitate the German government to meet its objectives of Platform Industrie 4.0 initiative (Zhong et al. 2017).

Few typical features such as learning, reasoning and even acting are effectively provided to the intelligent manufacturing environment by artificial intelligence. The human involvement could be reduced with the employability of artificial intelligence technologies. As, for instance, the automatic arrangement of product and material composition is accomplished through typical features of artificial intelligence. Another is that of real-time control and monitoring of the various manufacturing processes (Koren et al. 2017). With the concept of Industry 4.0 gathering pace, underpinning technologies such as autonomous sensing, intelligent decision-making, intelligent learning analysis and intelligent interconnecting have been realized by various manufacturing enterprises. The intelligent scheduling system has been employed that has aided in the scheduling of jobs on the basis of artificial techniques. The system then can be provided to other end users through Internet-enabled platform (Barbosa et al. 2015).

1.3 Key Techniques

Key technologies that are used in the realm of intelligent manufacturing such as Internet of Things, information and communication technology, big data analytics, cloud computing and cyber-physical systems have been discussed in this section.

1.3.1 Information and Communication Technology

Information and communication technology (ICT) is an extended version of information technology (IT) wherein a unified system comprising of communications and telecommunications enables to store, manipulate and transmit the data (Hashim 2007). ICT framework encompasses in itself a wide range of techniques related to computer science and signal processing. These include audio–visual systems, middleware, and wireless system. Information transmission is heavily relied on different types of electronic media such as wireless communication. This aspect plays a crucial role in intelligent manufacturing where various operations of production and decision-making rely on the information or data. ICT has resulted in encouraging impact on the plant managers or the workers associated with the task in the sense that it has allowed for enhanced autonomy and control (Bloom et al. 2014). For instance, the manufacturing sector in Europe has been found to be more competent with the adaptation of the ICT framework. In the backdrop of the ICT framework, the business environment was reported to be more flexible, agile and productive.

ICT has the potential ability to aid SME in responding satisfactorily to the dynamic markets. Handling of data or information has been managed with ease through the ICT technologies and has ultimately resulted in cost reduction as well as enhanced compliance of the clients (Colin et al. 2015). ICT has, therefore, become the foundation of manufacturing systems in recent times, which has been possible through the digital devices that have access through Internet-based networks. Customized designs that have the adaptability, rapid production of the customized design and delivery of customized products are some of the key features of the ICT framework that have been aided by digital platforms, virtual technologies, simulation and the related modelling tools and last but not the least the presentation and visualization tools (Ketteni et al. 2015).

Few key sectors that have exploited the benefits of ICT framework are medical, telemedicine, telecommunication, manufacturing, tourism and education. The history of the application of ICT is relatively longer in comparison to other technologies. The main reason being that ICT is an expanded version of the existing and several decade old computer technologies. ICT has been integrated with other key enabling technologies such as Internet of Things, cloud computing, etc. Significant improvements in cases of the real world have been reported with the usage of ICT technologies. Manufacturing companies now have a greater inclination for ICT framework to solve their issues of productivity and designs. The inclination will continue to increase in the context of Industry 4.0 owing to the umpteen benefits associated with the ICT.

1.3.2 Big Data Analytics

Data and information have become relatively easier to access and therefore ubiquitous for many industries as a result of greater interest towards data networks and technologies such as Internet of Things. Accessibility of data at ease has given rise to the concept of big data (Manyika et al. 2011). The origin of big data lies in various channels, networks, audio and video devices, web, social media, applications of transaction and log files (Rich 2012). The result has been the birth of big data environment in the sector of manufacturing. The collection of data has been streamlined through the advancements in Internet of Things technology. However, the main challenge lies into accurately process the collected data so that the right information for the desired purpose can be provided at an appropriate time (Lee et al. 2013). Conventional data analysis software finds it difficult to process a large and complex dataset in the environment of big data (Barton et al. 2012). Hence, it is quintessential to employ advanced analytics techniques in cases of manufacturing setups with a large amount of operational as well as shop-floor data. Employment of such techniques can aid in unraveling of hidden patterns, trends of market, and preferences of the customer, unknown correlations and the related useful information.

Surveys have been conducted to research on the return on investment that the industries have garnered with the implementation of big data analytic technologies (BDA). It has been revealed that around 15–20% increased return on investment could be achieved by the industries with the implementation of BDA techniques (Zhong et al. 2017). Customer engagements as well as their satisfaction could well be improved with the inclusion of customer relationship management data into analytics (Wamba et al. 2015). For instance, an automobile company can take into consideration customer satisfaction while launching a new car through data mining techniques, wherein the data obtained through user feedback. The productivity as well as the competitiveness of the manufacturing company can be truly realized through an in-depth analysis of data obtained from different machines and processes (Agarwal and Weill 2012). For instance, in pharmaceutical industries, a large number of parameters must be monitored in order to ensure the quality of production. Critical parameters can be discovered by a manufacturer through the processing of big data. Variables that affect critically the production activities and have a significant impact on product quality can also be identified through BDA techniques (Brown et al. 2011).

With the maturity of BDA technology, manufacturing firms such as General Electric have been proactively employing BDA for optimization of processes related to production and maintenance. There are now umpteen references for nascent manufacturers that are inclined towards the implementation of BDA to reap the benefits of optimized produce.

1.3.3 Cloud Computing

In the cloud computing environment, computational goods and services are delivered through the aid of technologies that support visualization and scalability. The dissemination is processed over the Internet (Armbrust et al. 2010). Scalability feature aids the business owners to start with something small and then motivate them to invest in more resources in accordance with the market or service demands (Zhang et al. 2010). Five characteristic features have been recommended for an ideal cloud by National Institute of Standards and Technology (NIST): measured service, rapid elasticity, and resource pooling, broad network access and on-demand self-service. Four deployment models make up this cloud model: private, community, hybrid and public. Furthermore, three delivery models are also encompassed by this cloud: infrastructure, platform and software, all as services (Mell and Grance 2009). Different types and sizes of organizations have been able to increase their capacity with the employability of cloud computing services. Manufacturing enterprises were able to do so without investing in new licensing software, personnel training and licensed software (Zhong et al. 2017).

There are certain challenges that are critically impacting the reliability of cloud computing services (Tan and Ai 2011). A number of studies have been conducted by the researchers to identify and therefore to classify the different issues related to the cloud computing services. Security and privacy-related challenges have been identified as the most significant factor affecting the reliability of cloud computing services (de Chaves et al. 2011; Banyal et al. 2013). Some of the other related issues that have been identified were: data management and allocation of resources (Maguluri et al. 2012; Sharkh et al. 2013), availability and scalability (Moreno-Vozmediano et al. 2013), balancing of load (Zhong et al. 2017) compatibility of clouds and their migration (Khajeh-Hosseini et al. 2010) and communication between different clouds (Petcu 2011).

Cloud computing has been often regarded as 'fifth utility' after electricity, water, telephone and gas. This has been made possible with the current advances in ICT (Buyya et al. 2009) and an array of research conducted in the domain of cloud computing (Yang and Tate 2009). Some of the typical applications of cloud computing include education, health care, transportation, manufacturing and so on. All the applications that can be run using a normal computer can be performed satisfactorily by the framework of cloud computing given that right middleware has been adopted. A cloud system can process varied programs ranging from word processing to a few customized business programs for an organization. Some of the major benefits associated with cloud computing are greater flexibility, elasticity, reduced cost and optimal utilization of resources. These advantages has ultimately resulted in enhanced competitiveness of the market.

1.3.4 Cyber-Physical System

Within the cyber-physical system (CPS), physical objects and different software platforms are integrated that helps the interaction between different components possible and hence exchange the relevant information (Lee 2008). A number of interdisciplinary subjects, for instance, mechanical engineering, mechatronics, manufacturing technology, cybernetics, design, computer science and process design are involved within a CPS framework. Embedded systems is one of the key technical methods that aids in high and precise coordination between physical objects and computational services (Tan et al. 2008). Networked interactions are encompassed within a CPS-enabled system. These interactions are designed with physical input and output and are integrated with cyber services such as computational capacities and control algorithms. A number of sensors are therefore employed for the successful CPS-enabled systems. Examples of such sensors include force sensors, light sensors, touch screens, etc., and therefore aids in achieving different purposes. However, it is time consuming as well as costly to integrate the several subsystems and also challenging to keep the entire system operating and functioning with consistency. Other challenges associated with the CPS systems are designing secure and high confidence systems and therefore the control techniques (Derler et al. 2012).

There has been a number of instances of industries that have initiated projects concerning the CPS-enabled systems. Festo Motion Terminal is one such example of a standardized platform that employs an intelligent fusion of mechanics, software application, electronics and the related sensors and other multisensory devices (Zhong et al. 2017). The subsystems are associated with self-adjusting and self-adapting characteristics through the aid of digital pneumatics. Sensor-based autonomous systems that are communication-enabled is the form of application for CPS-enabled framework that has been employed in diverse fields. Information from the environment has been managed and controlled centrally with the help of wireless sensor networks (Ali et al. 2015).

CPSs have gained tremendous popularity since their inception amongst the industry and academia. Employing the CPS-enabled framework, firms have been able to maintain global competitiveness in the market. There has been rapid pace advancements and developments to design CPS owing to their promising characteristic in maintaining global competitiveness in the market. Countries have been investing heavily in this interesting domain. Requirements, challenges and opportunities in design and development of CPSs have been rightly identified by the engineers, computer scientists and industrial experts. There has been a significant impact of employing CPSs in various fields such as autonomous vehicles, power distribution, intelligent manufacturing and civil structures.

1.3.5 Internet of Things

Internet of Things (IoT) is a digital environment wherein different manufacturing resources and objects are embedded with actuators, digital devices and sensors. The embedded objects are then connected together and therefore data is collected and exchanged between them (Xia et al. 2012). Thus, the IoT framework aids in object-to-object communication and sharing of data through the enhanced connectivity between physical objects, services and systems. IoT environment has aided various industries in achieving automation for various purposes such as machining, heating, space lighting and remote monitoring. One of the key characteristics of IoT framework is to create smart objects through automatic identification technology. One of the historical examples of IoT framework application includes that of an Internet-connected appliance which was applied by researchers at Carnegie Mellon University to a coke machine (Farooq et al. 2015). With time and advancements, IoT has grown in stature and is now regarded as a larger convergence of certain cutting edge technologies including machine learning, data analytics and wireless standards (Da Xu et al. 2014). This signifies that traditional areas that are linked to our daily lives will greatly be affected by IoT-based technologies.

RFID technology is one such example wherein it is believed that around twenty billion devices will be interconnected and thereby making full use of the capability of RFID technology by 2020 (Lund et al. 2014). Such technologies have impacted various industries mainly the manufacturing enterprises. The RFID technology has been employed for the identification and differentiation of objects in production shop floors, warehouses, disposal/recycle stages, logistics, retailers, distribution centres, etc. The objects identified have sensing capabilities and therefore can interact with one another, creating a huge amount of data. Such objects are embedded with the logic and procedures such as that related to manufacturing and other related logics when equipped with RFID tags (Guo et al. 2015). Therefore, RFID aids the users to fulfil their requirements and operations as desired through the smart objects. Realtime management of the production schedule is also achieved using the captured data.

Many countries are working in collaboration to employ IoT for achieving their desired projects. Countries such as France, China and India have been investing in IoT-based framework to meet the desired objectives. The collaborations also aid in addressing global issues besides enhancing the development of IoT techniques. Collaborations are quintessential for successful adoption and implementation of cutting edge technologies such as that of IoT.

1.4 Intelligent Manufacturing Platforms

Intelligent manufacturing technology has gained widespread acceptability amongst the industries and therefore cloud computing platforms are being developed by industries for successful implementation of intelligent manufacturing. Most of these cloud computing platforms are built based on IoT conceptual framework and therefore investments to the tune of $60 trillion have been estimated in this domain within a span of 15 years. The estimation is also that by the end of 2020 more than fifty billion devices will be accessible over the Internet. The applicability and benefits of IoT for the manufacturing sector have been evidenced through the development of a number of software platforms such as Predix software platform developed by General Electric (GE). Such platforms have the potential capability to transform a physical object into a computerized model, which is usually is referred to as "digital twins".

1.4.1 GE: Predix

GE's Predix is one of the software platforms for the implementation of intelligent manufacturing systems. It comprehensively monitors and controls various physical devices and the related systems over the Internet network. Predix Machines, Predix Cloud, Predix Edge are the key components of the Predix platform. The cloud model used in the Predix platform is the edge-to-cloud deployment mode. The architecture of Predix platform has been successfully implemented to various industrial applications.

Predix was an outcome of General Electric's (GE) own practice. GE was faced with a challenge to build digital twins for the sole motive of controlling and monitoring of heavy machines such as turbines. The company then came up with the solution and developed edges and twins, which were then successfully used through the Internet. The developed platform was then opened to the public, and as a result the third-party developers as well as original equipment manufacturers were able to develop twin virtual models for different systems.

Cloud Foundry is the core of Predix platform that provides an open source PaaS. The Predix platform supports many languages and tools for programming. Micro-services architecture of Predix is unique in itself and the third-party developers have found it easier to build, test and implement the various models for industrial applications owing to the modern development-and-operations environment. With the continual development process, there are now more than five lakh digital twins based on the Predix platform.

The Predix platform tends to collect data related to the manufacturing activities, which is humongous in volumes and therefore it becomes impossible for any public cloud to handle the data. Predix, therefore, aids industries in processing and analysing the activities related to manufacturing tasks and therefore enable them to make appropriate decisions. The companies can make such decisions in real time that ultimately results in improved business operations. Some of the major applications of Predix are: optimization of operations, asset performance, connected products, field-force management and application performance management.

Predix platform is in its nascent stage of development and ongoing efforts are in continual process for its deployment. It is required to further enhance the deep learning as well as the artificial intelligence framework.

1.4.2   PTC: ThingWorx

ThingWorx was acquired by PTC in 2014 and which then integrated it with its Internet-based PLM program, resulting in one of the major industrial Internet of things platform. Since its integration, ThingWorx has been proactive in the deployment of intelligent manufacturing technology. It encompasses in it several elements such as ThingWorx Utilities, ThingWorx Analytics, ThingWorx Industrial Connectivity and ThingWorx Studio. All the aforementioned elements work in tandem under the aegis of ThingWorx Foundation.

Connectivity between the different elements with end-to-end security is provided by the ThingWorx Foundation. This connective network aids the users in the creation and deployment of various industrial applications throughout the IoT environment. Edge, Connection Services and Core are the three fundamental functions provided by ThingWorx.

Model-drive and ground-up approach were used to develop the ThingWorx platform. A number of drag-and-drop tools have been incorporated with the platform that enables the end users to address the desired application. A number of algorithms have been incorporated within the platform that aids in proper analysis and presentation of data.

1.4.3   Siemens: Smart Factory

One of the leading concepts within the realm of Industry 4.0 is that of the smart factory. Smart factory technology consists of two levels: first level focussing on the shop floor and the other level being targeted for the production system. In the first level, the devices related to production are integrated using wireless or wired systems. The integration is such that there are no isolated production devices. Sensors, transducers and the related controllers aid in the collection of data in real time. The collected data provides information related to the working conditions of devices and the related environmental parameters. The information aids the human operators in precise monitoring, and control of the related manufacturing systems.

In its second level, the smart factory is accommodated with the so-called digital twin, i.e. a full digital factory model for the production system. Sensors, programmable logic controllers, data acquisition systems and other controllers and communication devices connect the digitized factory model to the product life cycle management system. Effective monitoring and control of the events in real time are made possible by the real-time floor conditions reflected by the digital twin.

Instosite software was rolled out by Siemens to facilitate effective communication between the digital twins. The production and manufacturing related information is shared by this cloud-based application. Application provides for precise mapping of virtual factories located at various locations which ultimately results in sharing of data from various digital factories amongst engineers and managers. The changes desired are uploaded automatically to the related IT systems and thereby enhancing the productivity of the manufacturing system.

1.5 Predictive Analytics for Intelligent Manufacturing

To ensure the successful practical implementation of intelligent manufacturing technologies, the role of predictive analytics is of utmost importance. Various tech giants such as Google, Intel and Microsoft are making flawless efforts in development of predictive analytic technologies.

1.5.1 Cloud ML Platform from Google

Since long, Google has been involved in the development of Machine learning and Artificial Intelligence technologies. Google has developed the Cloud ML platform for disbursing the cloud-based machine learning services. Enhanced performance characteristics as well as precision has been reported with the machine learning algorithms based on Internet platforms. Capabilities such as image analysis, speech recognition, and text analysis are some of the unique features exhibited by Google artificial intelligence platforms. TensorFlow is one of the newly added component to the machine learning platform that has enhanced its industrial application. This open source software library processes data and has a data flow graph structure wherein the mathematical operations are represented by the nodes and data arrays represented with the edges. The developed technology has been successfully implemented for uncertain manufacturing environments.

1.5.2 Azure from Microsoft

Microsoft has developed an Azure framework that provides integrated cloud services. The Azure framework consists of integrated cloud services that aid in the creation of machine learning tools through ready-to-use algorithms. An interconnected processor can be used to deploy the developed application.

HDInsight and R server are some of the other key elements of the Azure platform that aids in the processing of data. Majority of applications of Azure platform caters to the commercial sector, for instance management of web services. Their applicability to the manufacturing domain still needs to be explored.

1.5.3 Cloud Machine Learning Platform from Google

Different deep machine learning and artificial intelligence platforms are being built by Google. Google Cloud Machine Learning platform provides for cloud-based machine learning services. The accuracy of this net-based framework has been reported to be excellent. Powerful analysis if text, image analysis potentiality, and speech recognition are the uniqueness of artificial intelligence platforms from Google.

TensorFlow has been recently added by Google to explore the capabilities of its machine learning platforms in the industrial sector. The software platform aids in the processing of large volumes of data even for a situation that characterizes itself of higher levels of uncertainty.

1.6 Conclusion

Manufacturing research and advancements have opened frontiers for a new industrial revolution. This new paradigm shift is referred to as Industry 4.0. Many governments across the globe are working to strengthen their manufacturing base and therefore improve their market shares. Key enabling digital technologies such as cyber-physical systems, Internet of Things, big data and cloud computing are aiding the industries and manufacturing units to shift towards a new industrial paradigm. Industrial platforms are being developed for the successful implementation of the new ecosystem for manufacturing.

References

R. Agarwal, P. Weill, The benefits of combining data with empathy. MIT Sloan Manag. Rev. **54**(1), 35 (2012)

S. Ali, S. Qaisar, H. Saeed, M. Khan, M. Naeem, A. Anpalagan, Network challenges for cyber physical systems with tiny wireless devices: a case study on reliable pipeline condition monitoring. Sensors **15**(4), 7172–7205 (2015)

M. Armbrust, A. Fox, R. Griffith, A.D. Joseph, R. Katz, A. Konwinski, G. Lee, D. Patterson, A. Rabkin, I. Stoica, M. Zaharia, A view of cloud computing. Commun. ACM **53**(4), 50–58 (2010)

R.K. Banyal, P. Jain, V.K. Jain, Multi-factor authentication framework for cloud computing, in *2013 Fifth International Conference on Computational Intelligence, Modelling and Simulation (CIMSim)* (IEEE, 2013), pp. 105–110

D. Barton, D. Court, Making advanced analytics work for you. Harvard Bus. Rev. **90**(10), 78–83 (2012)

J. Barbosa, P. Leitão, E. Adam, D. Trentesaux, Dynamic self-organization in holonic multi-agent manufacturing systems: the ADACOR evolution. Comput. Ind. **66**, 99–111 (2015)

N. Bloom, L. Garicano, R. Sadun, J. Van Reenen, The distinct effects of information technology and communication technology on firm organization. Manag. Sci. **60**(12), 2859–2885 (2014)

Z. Bi, L. Da Xu, C. Wang, Internet of things for enterprise systems of modern manufacturing. IEEE Trans. Ind. Inf. **10**(2), 1537–1546 (2014)

B. Brown, M. Chui, J. Manyika, Are you ready for the era of 'big data'. McKinsey Q. **4**(1), 24–35 (2011)

R. Buyya, C.S. Yeo, S. Venugopal, J. Broberg, I. Brandic, Cloud computing and emerging IT platforms: vision, hype, and reality for delivering computing as the 5th utility. Future Gener. Comput. Syst. **25**(6), 599–616 (2009)

T. Chen, Y.C. Wang, Z. Lin, Predictive distant operation and virtual control of computer numerical control machines. J Intell. Manuf. **28**(5), 1061–1077 (2017)

M. Colin, R. Galindo, O. Hernández, Information and communication technology as a key strategy for efficient supply chain management in manufacturing SMEs. Procedia Comput. Sci. **55**, 833–842 (2015)

L. Da Xu, W. He, S. Li, Internet of things in industries: a survey. IEEE Trans. Ind. Inf. **10**(4), 2233–2243 (2014)

Q. Dai, R. Zhong, G.Q. Huang, T. Qu, T. Zhang, T.Y. Luo, Radio frequency identification-enabled real-time manufacturing execution system: a case study in an automotive part manufacturer. Int. J. Comput. Integr. Manuf. **25**(1), 51–65 (2012)

J. Davis, T. Edgar, J. Porter, J. Bernaden, M. Sarli, Smart manufacturing, manufacturing intelligence and demand-dynamic performance. Comput. Chem. Eng. **47**, 145–156 (2012)

P. Derler, E.A. Lee, A.S. Vincentelli, Modeling cyber–physical systems. Proc. IEEE **100**(1), 13–28 (2012)

S.A. de Chaves, B. Carlos, C.M. Westphall, G.A. Gerônimo, Customer security concerns in cloud computing, in *Proceedings of the Tenth International Conference on Networks (ICN)* (2011), pp. 7–11

M.U. Farooq, M. Waseem, S. Mazhar, A. Khairi, T. Kamal, A review on Internet of Things (IoT). Int. J. Comput. Appl. **113**(1), 1–7 (2015)

A.B. Feeney, S.P. Frechette, V. Srinivasan, A portrait of an ISO STEP tolerancing standard as an enabler of smart manufacturing systems. J. Comput. Inf. Sci. Eng. **15**(2), 021001 (2015)

Z. Guo, E. Ngai, C. Yang, X. Liang, An RFID-based intelligent decision support system architecture for production monitoring and scheduling in a distributed manufacturing environment. Int. J. Prod. Econ. **159**, 16–28 (2015)

J. Hashim, Information communication technology (ICT) adoption among SME owners in Malaysia. Int. J. Bus. Inf. **2**(2) (2007)

G.Q. Huang, Y.F. Zhang, X. Chen, S.T. Newman, RFID-enabled real-time wireless manufacturing for adaptive assembly planning and control. J. Intell. Manuf. **19**(6), 701–713 (2008)

E. Ketteni, C. Kottaridi, T.P. Mamuneas, Information and communication technology and foreign direct investment: interactions and contributions to economic growth. Empir. Econ. **48**(4), 1525–1539 (2015)

A. Khajeh-Hosseini, D. Greenwood, I. Sommerville, Cloud migration: a case study of migrating an enterprise it system to IAAS, in *2010 IEEE 3rd International Conference on Cloud Computing (CLOUD)* (IEEE, 2010), pp. 450–457

Y. Koren, W. Wang, X. Gu, Value creation through design for scalability of reconfigurable manufacturing systems. Int. J. Prod. Res. **55**(5), 1227–1242 (2017)

A. Kusiak, *Intelligent Manufacturing Systems* (Prentice Hall Press, NJ, 1990), p. 448

E.A. Lee, Cyber physical systems: design challenges, in *11th IEEE Symposium on Object Oriented Real-Time Distributed Computing (ISORC)* (IEEE, 2008), pp. 363–369

J. Lee, E. Lapira, B. Bagheri, H.A. Kao, Recent advances and trends in predictive manufacturing systems in big data environment. Manuf. Lett. **1**(1), 38–41 (2013)

J. Lee, B. Bagheri, H.A. Kao, A cyber-physical systems architecture for industry 4.0-based manufacturing systems. Manuf. Lett. **3**, 18–23 (2015)

W.N. Liu, L.J. Zheng, D.H. Sun, X.Y. Liao, M. Zhao, J.M. Su, Y.X. Liu, RFID-enabled real-time production management system for Loncin motorcycle assembly line. Int. J. Comput. Integr. Manuf. **25**(1), 86–99 (2012)

B.H. Li, B.C. Hou, W.T. Yu, X.B. Lu, C.W. Yang, Applications of artificial intelligence in intelligent manufacturing: a review. Front. Inf. Technol. Electron. Eng. **18**(1), 86–96 (2017)

B.H. Li, L. Zhang, S.L. Wang, F. Tao, J.W. Cao, X.D. Jiang, X. Song, X.D. Chai, Cloud manufacturing: a new service-oriented networked manufacturing model. Comput. Integr. Manuf. Syst. **16**(1), 1–7 (2010)

Y. Liu, X. Xu, L. Zhang, L. Wang, R.Y. Zhong, Workload-based multi-task scheduling in cloud manufacturing. Robot. Comput.-Integr. Manuf. **45**, 3–20 (2017)

B.H. Lu, R.J. Bateman, K. Cheng, RFID enabled manufacturing: fundamentals, methodology and applications. Int. J. Agile Syst. Manag. **1**(1), 73–92 (2006)

D. Lund, C. MacGillivray, V. Turner, M. Morales, Worldwide and regional Internet of Things (IoT) 2014–2020 forecast: a virtuous circle of proven value and demand. International Data Corporation (IDC), Technical Report, 1 (2014)

J. Manyika, M. Chui, B. Brown, J. Bughin, R. Dobbs, C. Roxburgh, A.H. Byers, Big data: the next frontier for innovation, competition, and productivity (2011)

S.T. Maguluri, R. Srikant, L. Ying, March. Stochastic models of load balancing and scheduling in cloud computing clusters, in *2012 Proceedings IEEE INFOCOM* (IEEE, 2012), pp. 702–710

P. Mell, T. Grance, The NIST definition of cloud computing. Natl. Inst. Stand. Technol. **53**(6), 50 (2009)

D. McFarlane, S. Sarma, J.L. Chirn, C. Wong, K. Ashton, Auto ID systems and intelligent manufacturing control. Eng. Appl. Artif. Intell. **16**(4), 365–376 (2003)

R. Moreno-Vozmediano, R.S. Montero, I.M. Llorente, Key challenges in cloud computing: enabling the future internet of services. IEEE Internet Comput. **17**(4), 18–25 (2013)

D. Petcu, Portability and interoperability between clouds: challenges and case study, in *European Conference on a Service-Based Internet* (Springer, Berlin, Heidelberg, 2011), pp. 62–74

S. Rich, Big data is a 'new natural resource'. Retrieved, 17 (2012), p. 2016

T. Qu, H.D. Yang, G.Q. Huang, Y.F. Zhang, H. Luo, W. Qin, A case of implementing RFID-based real-time shop-floor material management for household electrical appliance manufacturers. J. Intell. Manuf. **23**(6), 2343–2356 (2012)

M.A. Sharkh, M. Jammal, A. Shami, A. Ouda, Resource allocation in a network-based cloud computing environment: design challenges. IEEE Commun. Mag. **51**(11), 46–52 (2013)

W. Shen, D.H. Norrie, Agent-based systems for intelligent manufacturing: a state-of-the-art survey. Knowl. Inf. Syst. **1**(2), 129–156 (1999)

F. Tao, Y. Zuo, L. Da Xu, L. Zhang, IoT-based intelligent perception and access of manufacturing resource toward cloud manufacturing. IEEE Trans. Industrial Informatics **10**(2), 1547–1557 (2014a)

F. Tao, Y. Cheng, L. Da Xu, L. Zhang, B.H. Li, CCIoT-CMfg: cloud computing and internet of things-based cloud manufacturing service system. IEEE Trans. Ind. Inf. **10**(2), 1435–1442 (2014b)

X. Tan, B. Ai, The issues of cloud computing security in high-speed railway, in *2011 International Conference on Electronic and Mechanical Engineering and Information Technology (EMEIT)*, vol. 8 (IEEE, 2011), pp. 4358–4363

Y. Tan, S. Goddard, L.C. Perez, A prototype architecture for cyber-physical systems. ACM Sigbed Rev. **5**(1), 26 (2008)

J. Wan, S. Tang, D. Li, S. Wang, C. Liu, H. Abbas, A.V. Vasilakos, A manufacturing big data solution for active preventive maintenance. IEEE Trans. Ind. Inf. **13**(4), 2039–2047 (2017)

X.V. Wang, X.W. Xu, An interoperable solution for cloud manufacturing. Robot. Comput.-Integr. Manuf. **29**(4), 232–247 (2013)

S. Wang, J. Wan, D. Zhang, D. Li, C. Zhang, Towards smart factory for industry 4.0: a self-organized multi-agent system with big data based feedback and coordination. Comput. Netw. **101**, 158–168 (2016a)

S. Wang, J. Wan, D. Li, C. Zhang, Implementing smart factory of industrie 4.0: an outlook. Int. J. Distrib. Sensor Netw. **12**(1), 3159805 (2016b)

S.F. Wamba, S. Akter, A. Edwards, G. Chopin, D. Gnanzou, How 'big data' can make big impact: findings from a systematic review and a longitudinal case study. Int. J. Prod. Econ. **165**, 234–246 (2015)

D. Wu, M.J. Greer, D.W. Rosen, D. Schaefer, Cloud manufacturing: strategic vision and state-of-the-art. J. Manuf. Syst. **32**(4), 564–579 (2013)

F. Xia, L.T. Yang, L. Wang, A. Vinel, Internet of Things. Int. J. Commun Syst **25**(9), 1101–1102 (2012)

X. Xu, From cloud computing to cloud manufacturing. Robot. Comput.-Integr. Manuf. **28**(1), 75–86 (2012)

H. Yang, M. Tate, Where are we at with cloud computing?: a descriptive literature review, in *20th Australasian Conference on Information Systems* (2009), pp. 2–4

L. Zhang, Y. Luo, F. Tao, B.H. Li, L. Ren, X. Zhang, H. Guo, Y. Cheng, A. Hu, Y. Liu, Cloud manufacturing: a new manufacturing paradigm. Enterp. Inf. Syst. **8**(2), 167–187 (2014)

R.Y. Zhong, Q.Y. Dai, T. Qu, G.J. Hu, G.Q. Huang, RFID-enabled real-time manufacturing execution system for mass-customization production. Robot. Comput.-Integr. Manuf. **29**(2), 283–292 (2013a)

R.Y. Zhong, Z. Li, L.Y. Pang, Y. Pan, T. Qu, G.Q. Huang, RFID-enabled real-time advanced planning and scheduling shell for production decision making. Int. J. Comput. Integr. Manuf. **26**(7), 649–662 (2013b)

Q. Zhang, L. Cheng, R. Boutaba, Cloud computing: state-of-the-art and research challenges. J. Internet Serv. Appl. **1**(1), 7–18 (2010)

R.Y. Zhong, X. Xu, E. Klotz, S.T. Newman, Intelligent manufacturing in the context of industry 4.0: a review. Engineering **3**(5), 616–630 (2017)

Chapter 2
Process Planning in Era 4.0

2.1 Introduction

The advent of the Fourth Industrial Revolution and its continual progress has resulted in drastic changes in the associated professions. Personnel in the manufacturing domain are required to not only learn the new tasks but also get acquainted with the different high-tech gadgets (Gorecky et al. 2014). Furthermore, these personnel needs to ensure the correctness of the data generated by the machines linked in the network. Ensuring the correctness of the data is difficult given the amount of associated data. After ensuring the correctness of the data, decision-making and therefore predictive analysis is another important task (Waller and Fawcett 2013).

Acquiring of Industry 4.0 (I 4.0) practices by the developed countries is easier given the younger working force. However, for the developing countries where the working force is relatively older and therefore familiarization with the high-tech gadgets presents a hurdle (Mason and Lee 2006). With the relatively older task force, the education process is longer and the resistance to learn new technologies is much stronger (Lines et al. 2015). One of the newer professions is that of process planner which is getting a newer look in the era of I 4.0. Era I 4.0 involves the automation of working environment through the employability of various platforms such as cyber-physical systems and Internet of things. The associated data is processed online through the cloud computing framework (Wang et al. 2016). Due to the created automated machine network, machine-to-machine interactions have been truly realized. However, humans associated with the different production processes are getting eliminated. The major question that is certain in the realm of I 4.0, is whether the process planning can be automated? The knowledge requirements for the same should be realized and the required arrangements for its transfer to the human task force must also be realized.

Process planning in I 4.0 requires automated decision-making for sequencing and scheduling of operations, selection of primary processes and the related activities. Therefore, a system is required that can plan the process required as per the CAD model of the desired product. Furthermore, the system must be able to take any

K. Kumar et al., *Industry 4.0*, Manufacturing and Surface Engineering,
https://doi.org/10.1007/978-981-13-8165-2_2

modifications from the customer at a later stage. This is how process planning is to be recognized in the era of I 4.0. Implementation of artificial intelligence in process planning can aid in realizing the aforementioned process planning framework (Pedagopu and Kumar 2014).

In the present chapter, exploration will be done for the process planning in the era of I 4.0 environment. Some of the achievements in this regard will be highlighted such as that of CAPP and product planning software.

2.2 Traditional Process Planning and Process Planning I 4.0

The traditional approach of process planning in various small and medium enterprises is done on the basis of knowledge and experience of the personnel involved in the manufacturing activities. This personnel are experts in terms of the experienced gain and not by the knowledge they associate with. They learn through the working experience and not by any modern learning systems. Therefore, one of the major hurdles in the implementation of I 4.0 is their resistance to get hands-on experience of different modern systems. Furthermore, it becomes really difficult to convince such personnel to leave their traditional learning process and adapt to the changing high-tech modern systems (Cummings and Worley 2014). Companies also tend to resist to such adoption owing to the higher investments involved in purchasing of newer equipment and hence the digitization process.

However, due to the increased global competition, there exists a window for accepting solutions with characteristics of I 4.0. Characteristics feature of I 4.0 involves digitization and automation of every process and activities related to manufacturing. The big companies that motivate for research and development have slowly begun to accept the automation and digitization process and transform their working environment to I 4.0. The big companies do not have the hurdle of making heavy investments and are flexible enough to make themselves more competitive in the market. However, small and medium enterprises that have limitations of making big investments will be left behind in the global market as they will not be able to keep up with the change. Hence it becomes very critical for them to strategies themselves for implementing I 4.0 at the correct time.

In the realm of I 4.0, all the manufacturing resources are connected through data and communicate with one another by exchanging information. The correctness of the data is ensured through constant checks on quality and proper process control. The software systems have the capability to affect the business network of the company through direct communication with the needs of the supplier and small customers (Brettel et al. 2014). Products are no more traditional but are smart as they carry information and are able to directly link the customer feedback to the manufacturing system. The manufacturing system then uses the information gathered to optimize the product. Under such an environment of manufacturing, the customers are the protagonist for the company who effectively provide solutions and suggestions that aid the manufacturing unit to effectively carve out the customized products (Qin et al. 2016).

The efficiency of the whole supply chain is improved effectively through the employability of dynamic scheduling process. Software such as Structure Dynamic Control (SDC) has been used to effectively implement the dynamic scheduling process. However, a challenge still exists for determination of the optimal value of information desired to ensure successful operation of the different physical systems involved in the manufacturing system. Furthermore, it is difficult to decipher the information required to properly schedule and plan and reschedule the different activities for execution at the control stage (Ivanov et al. 2016). Digitization and automation of process planning exist for spot welding. Once the required information is deciphered, the same is transferred to software that on the basis of the inputs provides the required output (Andersson et al. 2016). The aforementioned arrangement can be realized truly if the manufacturing company has only one kind of production technology. The major challenge transpires when the manufacturing company has more than one kind of production technology. One can always suggest the solution on the basis of basic principle, however, will the solution be implemented in case the processes are more complicated and hence the data gets bigger? Physical manufacturing processes are now being virtualized and now referred to as cloud manufacturing. In this environment of virtualization, the virtual reality and hence the communication between various manufacturing processes have replaced the hardware part (Thames and Schaefer 2016). The products are now referred to as 'smart product' and are now being used to obtain the suggestions and feedback directly from the customer (Gibbert et al. 2002). However, the main question arises as to whether the product can carry the knowledge and information desired for suitable process planning and scheduling. The very question lays the topic of great research amongst the research community involved directly or indirectly.

2.2.1 Software for Product Planning

The design and development phase of a smart product must consider to include the data of the following three stages: process planning, operation scheduling and operation sequencing. The major aim must be to collect as much data as possible so that a digitalized and virtual network of manufacturing machines and the product can be established directly. Furthermore, the customer can be directly involved in providing their valuable suggestions and feedback to the manufacturing unit. This will aid in enhanced efficiency of the whole supply chain and mass customization of products. Product planning software with its various components has been depicted in Fig. 2.1.

One of the used process planning software is CAPP that establishes a direct link between CAM and CAD software (Engelke 1987). CAPP has been used widely to aid in mass customization of the products. However, the generic approach is considered from the technical viewpoint for effective implementation of CAPP for mass customization. In the generative approach, it is required to obtain the product plan for each single product using the data from the database (Monostori et al. 2016). The knowledge gathered from the knowledge database can be used for sequencing and

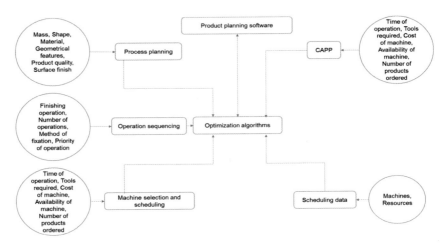

Fig. 2.1 The product planning software

scheduling of operations for the product. The CAD model of the product aids in providing the information for establishing the required technology. Any modifications required are defined using the new STEP model and therefore the 3D modelling software presents no boundaries and restrictions in process planning. That is why CAPP has gained widespread popularity amongst the industries in the Era I 4.0. However, the software prototype has many limitations and problems are being faced by the customers. For instance, linking of CAPP with other manufacturing units within a company has not been addressed. Furthermore, the continual progress and advancements in I 4.0 era demands for upgradation, wherein the software can be linked to various other manufacturing activities.

Therefore, efforts have been made to eliminate the drawbacks and link CAPP to product design and manufacturing phases. CAPP, therefore, now forms a part of product planning software that aids in process planning, scheduling operations and their sequencing. Its functional output is exploited at a later stage.

There are three important criteria associated with product planning software: process planning, operation sequencing and finally machine selection. The software for product planning collects information from different parts of the supply chain linked to the important manufacturing activities.

Within the main criteria, there are certain sub-criteria that need to fulfil for effective implementation of main criteria. Mass of the product, its material, shape, surface finish, geometrical features and the quality forms the sub-criteria for the process planning criteria. Mass of the product is linked both to the machining as well as logistic features. Material manipulations may be required during the manufacturing process. The mass constraint includes, for instance the possible size of the equipment or the size of the vehicle that can be accommodated in a hall or can transport the required personnel. The mass limitation can also include the number of robots that are required in the transportation of raw materials from the supplier or the number of units of the final product to be transported to the warehouses. Material on the other hand is related to the machining process besides the product and the input for its availability, price and delivery are linked and controlled through the supplier network. The archiving process is simplified through the shape sub-criterion. This is accomplished through the optimization of each and every product and therefore storing the optimized data for comparing with the older data. The geometry feature sub-criteria considers the tools and machines required to produce the required geometrical features on the product. A similar task is also performed by the surface quality sub-criteria, however, it also takes into account the cost and benefit calculations into consideration. Summary and the output received from the different sub-criteria are analysed by the product quality sub-criteria. The analysis is done by considering the valuable customer feedback and solutions. This aids in setting up the optimal quality of the product and the cost of the product.

The second criteria under product planning software is that of operation sequencing. There are numerous sub-criterion that can influence the operation sequencing criteria. However, the product planning software only takes into consideration the important ones that really influences the manufacturing processes, i.e. finishing operation, number of operations required, the fixation method and the priority of operation. In order to eliminate or reduce the wastage, roughing operation needs to be performed before finishing when the machining process is automatized. Number of operations is an important sub-criteria that makes a comparison between sequencing and processing. The fixation sub-criteria aims to fix the technology used in different operations. The same technology can be used for the performance of different operations. Therefore, the technologies are fixed such that the operations desired are followed consecutively one after the other and finally follows the forthcoming criteria. Piece manipulation is often demanded by the different fixation methods between two operations. Optimization and control aids in enhancing performance and productivity. This also results in reduced cost of the final product as well as shortening of the auxiliary times. The last important sub-criteria that fulfils the operation sequencing is that of operation prioritization. Often there are situations where one operation is required to be performed before another owing to the geometrical and organizational requirements. These requirements are sensed through the exchange of data between general system software and CAPP module.

The third major criteria for product planning software is that of machine selection and scheduling criteria. Related important sub-criteria includes: time of operation, tools required, quality required, cost of machine, availability of machine and number of products ordered. Selection of machine forms an important part of the product planning software, which is done through the information on the available machines in the manufacturing system. The software aids the product to itself to decide on the machines required for performing the desired operation on it. Time of operation encompasses technological time, auxiliary time and preparatory time. The total calculation results from the number of products required to be ordered which is the next sub-criteria within the machine selection and scheduling criteria. These are linked together such that decision is made in real time about the availability of machines. The knowledge database of the product planning software will have prior information about the different machines that can perform the same operation. Therefore, in accordance with the quantity of product desired the machines will be made available to the product under progress. The inbuilt advanced optimization algorithm aids the product planning software to calculate the optimal routes for inter-transport and therefore enable the manufacturing system to perform the operation with the overall objective of cost minimization and time of manufacturing.

The demand for product planning software has a higher demand in the era of I 4.0 as the software has been developed to recognize and analyse the different data associated with the various manufacturing processes. The analysis is done by it in real time resulting in the desired performance of the manufacturing system. Product planning software continuously upgrades its knowledge database and hence its optimization module. Different criteria are given importance in accordance with the product and customer feedback. There are two methods to develop a software package. One of the methods being that of predictive analysis which is accomplished using data mining. The predictive analysis approach uses many data from the database to create the patterns pertaining to the performance of the manufacturing systems and also recognize the different errors which ultimately leads to the prediction of the performance of the manufacturing system. The second method is that of decision support system that associates the weights to the different criteria and thereby obtains the solution as a final decision. The optimal approach would be to use both the methods, i.e. the usage of data mining to process a big amount of data and then to associate the importance weights to the identified criteria.

2.3 Possibility of Implementation

The new environment of manufacturing goods and services demands for digitization of every part of the manufacturing unit. As such continual research and developments have resulted in the digitization of product planning process. The conceptual framework has been defined completely and the industries that are ready for the I 4.0 environment are accepting this very change. With the wider acceptability, process planning platforms are being built and tested for practical implementation. The errors are identified and iteratively improvements have been carried to build the final version of the software. The readiness factor is taken into account for its final implementation. The companies that hesitate to implement such strategies are the ones that have hurdles to making bigger investments. The depth to which the implementation and changes can be accepted is defined after the readiness factor has been calculated. The implementation process takes time and therefore the new companies can easily implement the digitization processes. To suit the different requirements of a particular manufacturing unit, the product planning software can be built accordingly. However, the general version of the software could well do away with the problem of a customized software platform. It is also advisable to keep the software platform as open source so that the users can easily access the platform and define it in terms of their own needs. Secure software platforms is another area of great concern before the implementation process. The knowledge database must be organized as the information results in overall performance optimization and hence global advantage over the others in the competitive market.

The software of product planning is a part of the main software system that looks after the entire product development process. Software is directly linked to the other parts of the manufacturing unit through cloud computing and this allows for data manipulation in real time. There is direct communication with the different machines involved in the manufacturing process. The technical implementation of the software is easy as the software needs to collect the data from the databases and then make a proper analysis and therefore making the necessary decisions. The major problem is the database on which the software relies. Therefore, it is the development phase of the software that can take more time in comparison to the implementation phase. Developed software should form an integral part of the company's intellectual property.

2.4 Conclusion

Concept of I 4.0 should be developed and implemented by a company from the very beginning of its foundation. The human factor associated with manufacturing systems diminishes with the implementation of various advanced technologies, big data analytics and Internet of things. Also, the various related professions undergo drastic changes. One such profession is that of process planning. In the realm of I 4.0 every process and activities need to be automatized and so is the requirement to automatize process planning. This chapter has discussed process planning in the era I 4.0. Process planning has gradually transformed into product planning. With the conceptual framework of digitization and virtual reality, product planning has been implemented through software platforms. Customers are the protagonist for the whole manufacturing system and as such this has been realized truly through the various product and process planning software platforms. Valuable feedback and the optimization process have been linked directly through various process and product planning software. Their implementation has resulted in maximization of productivity, quality and mass customization with a lower final cost of the product. However, their implementation is still a topic of great concern given the financial as well as security reasons.

References

O. Andersson, D. Semere, A. Melander, M. Arvidsson, B. Lindberg, Digitalization of process planning of spot welding in body-in-white. Procedia CIRP **50**, 618–623 (2016)

M. Brettel, N. Friederichsen, M. Keller, M. Rosenberg, How virtualization, decentralization and network building change the manufacturing landscape: an industry 4.0 perspective. Int. J. Mech. Ind. Sci. Eng. **8**(1), 37–44 (2014)

T.G. Cummings, C.G. Worley, *Organization Development and Change* (Cengage Learning, 2014)

W.D. Engelke, *How to Integrate CAD/CAM Systems: Management and Technology* (CRC Press, 1987)

D. Gorecky, M. Schmitt, M. Loskyll, D. Zühlke, Human-machine-interaction in the industry 4.0 era, in *2014 12th IEEE International Conference on Industrial Informatics (INDIN)*, IEEE, July 2014, pp. 289–294

M. Gibbert, M. Leibold, G. Probst, Five styles of customer knowledge management, and how smart companies use them to create value. Eur. Manag. J. **20**(5), 459–469 (2002)

D. Ivanov, B. Sokolov, M. Ivanova, Schedule coordination in cyber-physical supply networks Industry 4.0. IFAC-PapersOnLine **49**(12), 839–844 (2016)

B.C. Lines, K.T. Sullivan, J.B. Smithwick, J. Mischung, Overcoming resistance to change in engineering and construction: change management factors for owner organizations. Int. J. Project Manage. **33**(5), 1170–1179 (2015)

A. Mason, R. Lee, Reform and support systems for the elderly in developing countries: capturing the second demographic dividend. Genus, 11–35 (2006)

L. Monostori, B. Kádár, T. Bauernhansl, S. Kondoh, S. Kumara, G. Reinhart, O. Sauer, G. Schuh, W. Sihn, K. Ueda, Cyber-physical systems in manufacturing. CIRP Ann. **65**(2), 621–641 (2016)

V.M. Pedagopu, M. Kumar, Integration of CAD/CAPP/CAM/CNC to augment the efficiency of CIM. Int. Rev. Appl. Eng. Res. **4**(2), 171–176 (2014)

J. Qin, Y. Liu, R. Grosvenor, A categorical framework of manufacturing for industry 4.0 and beyond. Procedia CIRP **52**, 173–178 (2016)

L. Thames, D. Schaefer, Software-defined cloud manufacturing for industry 4.0. Procedia CIRP **52**, 12–17 (2016)

M.A. Waller, S.E. Fawcett, Data science, predictive analytics, and big data: a revolution that will transform supply chain design and management. J. Bus. Logist. **34**(2), 77–84 (2013)

S. Wang, J. Wan, D. Zhang, D. Li, C. Zhang, Towards smart factory for industry 4.0: a self-organized multi-agent system with big data based feedback and coordination. Comput. Netw. **101**, 158–168 (2016)

Chapter 3
Requirements of Education and Qualification

3.1 Introduction

Technological innovations and changes have greatly affected the productivity of companies in the past. There have been various such paradigms and are referred to as industrial revolutions. The different paradigms are the First Industrial Revolution which focused on mechanization, Second Industrial Revolution focusing on electrical energy and the Third Industrial Revolution on electronics and automation (Lasi et al. 2014). The various industrial revolutions affected both the productivity of the companies as well as the labour market and hence the associated education system. As a result of these revolutions, the human factor, i.e. the associated jobs and professions disappeared. Presently the industries are experiencing Fourth Industrial Revolution with robotics and digitalization at the centre stage. The revolution is referred to as Industry 4.0 (I 4.0). In this era, it is expected that various professions will be replaced and many will undergo drastic changes. Coming of the new technologies have had a significant effect on the education of the people. The requirement calls for the highly skilled and qualified personnel that have the potential ability to control the new technologies. To meet the educational qualifications, industries are required to collaborate immensely with various universities across the globe (Baygin et al. 2016). The main vision of Industry 4.0, i.e. to establish smart factories can be realized with the involvement of highly skilled personnel. Various factories will form a network by linking themselves through cyber-physical systems (CPS) (Lee et al. 2015). The know-how on the latest technologies such as Internet of Things, Internet of Services, etc., will aid in establishing efficient and secure connections between machines, humans and machines and even humans. However, these connections will result in the generation of large quantum of data. Therefore, it also becomes essential to educate people on big data that can help in the prediction of possible failures and hence rectify the same under real-time situations (Richert et al. 2016). In the absence of 4.0 environment, various machines are operated and monitored by various operators. However, in I 4.0 environment, the manual monitoring is being replaced with prognostics monitoring system (Lee et al. 2014). Production processes in the Era 4.0

© The Author(s), under exclusive licence to Springer Nature Singapore Pte Ltd. 2019
K. Kumar et al., *Industry 4.0*, Manufacturing and Surface Engineering,
https://doi.org/10.1007/978-981-13-8165-2_3

are more effective and flexible and therefore meet the changing customer demand for goods and services as and when required. The companies investing big in the digitization process aims to create smart objects that can aid in the optimization of the product (Nelles et al. 2016). However, for the efficient implementation of flexible planning and management systems, timely analysis of the obtained data is critical. The data generated consists of classified information and this calls for educating personnel on cybersecurity so that potential data leaks can be prevented.

Owing to the different new professions and the changing existing ones, the role of humans is very much critical to realize manufacturing in the era of I 4.0. The development and successful implementation of the smart factories will depend to a great extent on the skills and qualification of the workforce (Gehrke et al. 2015). It is required for the human resource management to be focused on the development of a qualified workforce (Armstrong and Taylor 2014). Besides dealing with the selection of the workforce, and their dismissal, human resource management must also take care to educate and provide learning and training to its employees from time to time (Becker 2013). The competency level to judge the skill of the workforce can be aggregated from four main competencies: Personal, Technical, Methodological and Social (Hecklau et al. 2016). Obviously, due to the employability of newer technologies, the requirements on qualification and skills for the employees will be higher than that at the present. Therefore, in the era of I 4.0, education and learning translates to Education 4.0 (E 4.0) from Education 3.0 (Harkins et al. 2008). E 4.0 is combination of real and the virtual world wherein, for instance the teaching and learning would be done through virtual glasses (Quint et al. 2015). There will be upgradation of the existing subject database, for instance information science will also encompass knowledge database on process management (Pfeiffer 2015). E 4.0 will ultimately result in the upgradation of knowledge, qualification and required staff training and therefore will form a key component of I 4.0 (Huba and Kozák 2016). Developed knowledge database could be transferred through the various virtual learning platforms. The teachers will adorn the new role of avatars of these virtual learning platforms (Anjarichert et al. 2016). Virtual platforms as a first will aid in the successful implementation of E 4.0 and therefore educating the employees. However, the implementation is incomplete without interfacing the augmented virtual world with the real world (Benešová and Tupa 2017; Rüßmann et al. 2015). However, the education system will be cost intensive. This limitation can be done away with the privatization of some institutions or with the companies collaborating and investing in higher education (Störmer et al. 2014).

3.2 Phases of I 4.0 Implementation and the Required Jobs

The successful implementation of I 4.0 cannot be done overnight, but it is a process that can take place gradually. Implementation of I 4.0 to small- and medium-scale enterprises needs to be done in several phases so that the change from third to I 4.0 takes place gradually.

Once the different phases of I 4.0 implementation have been determined, the required jobs would be easy to determine. Before the implementation of I 4.0, it is necessary to ensure that the target enterprise has already installed the basic information system which allows for bookkeeping, displaying of important indicators that allow for precise business management and administration of human resources. However, if the target enterprise does not have the basic information system in place then the implementation of I 4.0 should start from zero phase. Therefore, execution of zero phase with advanced information system forms the initialization of the implementation phase of I 4.0. Although, zero phase uncovers some of the basic problems, the bigger problems of lack of technological skills and the age of employees are uncovered and solved in the latter phases of I 4.0 implementation.

Broadly, there are four phases to implementation of I 4.0 (Benesova and Tupa 2017) the first phase is about digitally representing the business enterprise, the second phase is about horizontal integration that leads to the installation of automated machine, the third phase of implementation is about vertical integration wherein deployment of sophisticated methods for processing of data is accomplished and the fourth phase is about automated control of manufacturing processes and the whole supply chain.

The first phase of I 4.0 implementation phase comprises of introducing the advanced information system. This phase incepts after the basic information system has been ensured to be installed within the target enterprise. ERP and ERP II are examples of an advanced information system. Mapping of all resources related to the manufacturing and production process becomes easier with the introduction of advanced information system within an enterprise. Therefore, enterprises require engineers and workforce that are qualified enough to look after the mapping process. Even the second phase of I 4.0 implementation demands the services of process engineers. Digitization of the existing data in the basic information system is also done within this phase of I 4.0 implementation. The data model must be able to replicate the real-world scenario. The digitization process ensures the increased volume of stored data.

To handle the increased volume of generated data, cloud services and the related systems are demanded. Therefore to implement the various cloud services, specialists with IT knowledge are required. The cloud systems also require the services of integrators and system engineers. After the successful implementation of cloud systems, the IT specialists will adorn the role of IT maintenance engineer as they would be responsible for the maintenance of server and hardware. Besides the maintenance task, the engineers are required to support users in the new system.

The automated machines will be implemented in the second phase of I 4.0 implementation. The new automated machines results in the reincarnation of the production processes. This phase also requires the services of process engineers. The existing employees of the company can also undergo retraining for managing the new automated machines. Hence, some of the probable workers can become process engineers with upgradation of their education in the relevant field. The aforementioned strategy is convenient to an enterprise as the existing workforce is aware of the environment and the different practices and processes of the enterprise.

The retraining process could be adopted in two ways: complete retraining of all the existing and new workforce by the supplier of the machinery or by providing special training to few of the employees who then can retrain their counterparts. The retrained workforce can then manage the operations of the automated machines. The minor defects in the machines can be looked after by technicians trained to do so.

Automation of the manufacturing processes will promote mass customization. Given the global market, a company that can satisfy the customer needs will obviously have a competitive advantage. The customer demand changes with the market, which ultimately affects the flexibility of a company. The company that is more flexible relative to the others will be able to fulfil its customer needs. Besides having an effect on the flexibility of the company, the customer requirements also affect the variability of manufactured products. Therefore, to meet the flexibility as well as the variability, the company would avoid losing on the workforce it has.

The new automated machinery is required to have diagnostic as well as telemetric units so that this can be connected to the network. This aids in the collection and analysis of generated data which is very much necessary for realizing the true vision of I 4.0.

The third phase of I 4.0 implementation makes use of the data generated in the first and second phase. Sophisticated methods for processing of generated data are deployed in the third phase of I 4.0 implementation. The major challenge of implementation, i.e. data analysis pops up at this phase of I 4.0 implementation. Field of data analysis is complicated and comprehensive and therefore it is very difficult to retrain the existing workforce of an enterprise to improve their education database in this field. The secondary problem arises to involve business data analysts and not just data analysts to aid the enterprise in ultimately enhancing their business performance. However, the cost constraint in the hiring of the same is higher. The small and medium enterprises can outsource the workforce for such tasks. Small and medium enterprises will also seek for data analysts, who not only have the knowledge about the data analysis but also about the production process.

In large business enterprises, efficient collaboration between the various departments is required so that the production processes can be controlled effectively. However, if there is no proper link and collaboration, the data generated will have no significant impact on the performance of the business enterprise as a whole.

The fourth phase of I 4.0 implementation is dedicated to ensure cent percent autonomous manufacturing. Different personnel such as quality control engineers, process engineers, data analysts, maintenance engineers, machine operators are required to ensure cent percent implementation of this phase. When implemented, the production process will be self-monitoring and optimizing and hence the productivity will be maximum. Since mutual communication will be required between personnel and the machines in case of large companies with their enterprises all over the world, the employees are required to be trained on language skills. The effective communication skills will be required in particular by the logistics department that is mainly responsible for the management of transport activities. These activities include flows related to the material and financial information that ultimately satisfy customer requirements.

On looking at the retraining and modifications in the role of the personnel, it is obvious that the first two phases will be difficult to implement. Therefore, companies that are I 4.0 ready will have to pay extra effort to train and retrain their existing employees for successful implantation of I 4.0.

3.3 Qualifications for Personnel for I 4.0

Qualifications and skills required by various personnel in I 4.0 ready enterprise have been provided by researchers Andrea Benešová and Jiří Tupa (Benesova and Tupa 2017). The skills required for various IT personnel demands for knowledge of different manufacturing processes. Small teams of IT technicians are managed by specialists in their respective fields. The teams will help in supporting individual processes such as maintenance of server systems, network infrastructure to ensure the effectiveness of I 4.0 environment. The major role is that of PLC programmer who aids in programming and therefore automation of the manufacturing systems. Usage of industrial robots to relieve the labour of heavy physical tasks which is one of the major objectives of I 4.0. Precise programming is required for their commissioning to be used dynamically. Such related tasks are performed by the robot programming specialist. The flexibility of the manufacturing systems is ensured by the information systems and the related software platforms are disbursed by the software engineers. Valuable information necessary for optimization is stored within the information database and related systems. The stored data is analysed by the data analysts. Therefore, security and communications of the stored data are really critical to protect the companies' secrecy of manufacturing operation. Cybersecurity personnel takes care of any possible attacks from hackers on the information and manufacturing systems.

Maintenance of machining equipment will be looked after by electronics technicians. The electronic technician is required to possess skills related to hydraulics and electrics. Automation technician on the other hand handles the design and installation of actuators and other related mechanical parts in a machine. Production technician will look after the automated production-related processes.

From the requirements of different personnel for carrying out various operations, it is obvious that jobs will not reduce as expected but only the job roles will change in the era of I 4.0.

3.4 Conclusion

It is very essential for companies to identify correctly the phases in which they are currently before incepting to implement I 4.0. Furthermore, on implementation, the companies must have a proper vision for the future to manage and regularly upgrade the installed capacity. On the basis of generated data, the analysis should be carried out precisely and any modifications or errors reflected should be done away gradually. With the gradually changing environment, some of the exiting job roles will modify whereas few of the jobs will disappear in the market. It is always beneficial for a company to retain the existing employees as they are already aware of the current manufacturing processes. The retraining of existing employees to upgrade their skills and qualifications may pose certain problems initially. To complement the retrained employees, IT specialists will be required to hire who can look after the field of computing, data analysis and optimization algorithms.

The education curricula existing currently should be changed considering tertiary education. Hence, at first instance, there will be a lack of specialists in the market who have the requisite knowledge. Survey questionnaires could well help various educational institutes to decipher the digital behaviour of the students and hence their outlook towards I 4.0. It is necessary to uncover the relationship between student and digital devices such as 3D printing, Mixed Reality, Virtual Reality, etc. Understanding of the relationship will aid in structuring the knowledge base so that basic concepts relating to the new industrial environment could be easily understood by the students. This will ultimately result in the production of well-qualified workforce that can fulfil the very objective of I 4.0 implementation.

As far as restructuring is concerned, the traditional machines would be replaced with the automated ones which will require the services of production, process and service engineers. Therefore, as far as jobs are concerned, new roles will be created by retraining of the existing employees and some new jobs wherein the companies would be required to hire the new staff.

References

L.P. Anjarichert, K. Gross, K. Schuster, S. Jeschke, *Learning 4.0: Virtual Immersive Engineering Education*, vol. 2 (Digital Universities, 2015); Int. Best Pract. Appl. **2–3**, 51

M. Armstrong, S. Taylor, *Armstrong's Handbook of Human Resource Management Practice* (Kogan Page Publishers, 2014)

M. Baygin, H. Yetis, M. Karakose, E. Akin, An effect analysis of industry 4.0 to higher education, in *2016 15th International Conference on Information Technology Based Higher Education and Training (ITHET)*, IEEE, September 2016, pp. 1–4

M. Becker. *Personalentwicklung: Bildung, Förderung und Organisationsentwicklung in Theorie und Praxis.* Schäffer-Poeschel Verlag für Wirtschaft Steuern Recht GmbH (2013)

A. Benešová, J. Tupa, Requirements for education and qualification of people in industry 4.0. Procedia Manuf. **11**, 2195–2202 (2017)

L. Gehrke, A.T. Kühn, D. Rule, P. Moore, C. Bellmann, S. Siemes, D. Dawood, S. Lakshmi, J. Kulik, M. Standley, A discussion of qualifications and skills in the factory of the future: a german and american perspective. VDI/ASME Industry **4**, 1–28 (2015)

A.M. Harkins, Leapfrog principles and practices: core components of education 3.0 and 4.0. Futures Res. Q. **24**(1), 19–31 (2008)

F. Hecklau, M. Galeitzke, S. Flachs, H. Kohl, Holistic approach for human resource management in industry 4.0. Procedia CIRP **54**, 1–6 (2016)

M. Huba, Š. Kozák, From E-learning to industry 4.0, in *2016 International Conference on Emerging eLearning Technologies and Applications (ICETA)*, IEEE, November 2016, pp. 103–108

H. Lasi, P. Fettke, H.G. Kemper, T. Feld, M. Hoffmann, *Industry 4.0. Business & Information Systems Engineering*, vol. 6, no. 4 (2014) pp. 239–242

J. Lee, H.A. Kao, S. Yang, Service innovation and smart analytics for industry 4.0 and big data environment. Procedia Cirp **16**, 3–8 (2014)

J. Lee, B. Bagheri, H.A. Kao, A cyber-physical systems architecture for industry 4.0-based manufacturing systems. Manuf. Lett. **3**, 18–23 (2015)

J. Nelles, S. Kuz, A. Mertens, C.M. Schlick, Human-centered design of assistance systems for production planning and control: the role of the human in Industry 4.0, in *2016 IEEE International Conference on Industrial Technology (ICIT)*, IEEE, March 2016, pp. 2099–2104

S. Pfeiffer, *Effects of Industry 4.0 on Vocational Education and Training* (2015)

F. Quint, K. Mura, D. Gorecky, In-factory learning-qualification for the factory of the future. ACTA Univ. Cibiniensis **66**(1), 159–164 (2015)

A. Richert, M. Shehadeh, L. Plumanns, K. Groß, K. Schuster, S. Jeschke, Educating engineers for industry 4.0: virtual worlds and human-robot-teams: empirical studies towards a new educational age, in *Global Engineering Education Conference (EDUCON)*, IEEE, April 2016, pp. 142–149

M. Rüßmann, M. Lorenz, P. Gerbert, M. Waldner, J. Justus, P. Engel, M. Harnisch, *Industry 4.0: The Future of Productivity and Growth in Manufacturing Industries*, vol. 9 (Boston Consulting Group, 2015)

E. Störmer, C. Patscha, J. Prendergast, C. Daheim, M. Rhisiart, P. Glover, H. Beck, *The Future of Work: Jobs and Skills in 2030* (2014)

Chapter 4
Risk Management Implementation

4.1 Introduction

Industry 4.0 (I 4.0) era uses various integrated software platforms and operations to connect different machines. The very concept of I 4.0 was introduced in Germany with the pitch towards internet in manufacturing. The third industrial revolution focused on the electronics and IT infrastructure in manufacturing. The fourth industrial revolution will be engrossed with the linking of sub-components of the manufacturing and production processes by employability of Internet of Things (IoT). The term I 4.0 was first mentioned way back in 2011 at Hanover Fair and was defined as a term encompassing various technologies and conceptual framework of value chain organizations. With its advent, various new technologies such as Internet of Energy, Internet of People, Internet of Services and Cyber-Physical Systems have been created (Hermann et al. 2016; Lom et al. 2016). Various organizations are being surveyed and have been surveyed for acceptability of I 4.0 environment (Oesterreich and Teuteberg 2016). A positive outlook of various companies towards I 4.0 have been revealed through these surveys. The companies that will accept to implement I 4.0 will transform to digital avatars. The physical manufacturing devices will be at the core which will be augmented by digital interfaces and other innovative services. A digital industrial ecosystem will be established wherein various enterprise will work in tandem with their suppliers and customers.

However with the digital transformation in I 4.0 era through acceptance of various high end technologies, probability of new risks and their negative impacts on the companies are inevitable. A new potential threat to the inherent data of the companies lurch through the integration of various IT infrastructure. These risk includes cyber-attack, spyware, malware and loss of data that may ultimately effect significantly the various manufacturing processes. Therefore there is a need to develop and test the risk management framework. Hackers and cyber pirates will aim to attack the documentation and specifications related to manufacturing and maintenance activities.

Owing to the importance of risk management in I 4.0 era, the present chapter deals with the various aspects related to risk management and its effective implementation.

© The Author(s), under exclusive licence to Springer Nature Singapore Pte Ltd. 2019
K. Kumar et al., *Industry 4.0*, Manufacturing and Surface Engineering,
https://doi.org/10.1007/978-981-13-8165-2_4

However, the major hurdle to this assessment is related to the software criminality and hence the problem of data reliability. The present chapter therefore discusses on how the potential threats could be minimized through the implementation of risk management framework.

4.2 Review on Existing Literature

There has been exponential rise in the number of publications in the area of I 4.0 which signifies a greater interest amongst the researchers and scientists working in this area. Strategies have been devised and adopted by the countries all over the globe to implement the very conceptual framework of I 4.0. As for instance, a document on 'Initiative I 4.0' was released by the Czech Republic for streamlining the implementation of I 4.0. The required resources were allocated by adoption of the document. However, major challenge rises is the associated risk with the complex IT infrastructure. The aim of the present section is to present the context that clubs the aspects of risk management and I 4.0.

Cyber Physical Systems lay the foundation of fourth industrial revolution which are the key enabling technologies in this area. These technologies support in creation of components for intelligent manufacturing environment and hence the environment of new production systems. The Cyber Physical Systems are termed as Cyber Physical Production Systems within the context of industrial framework. The integration of these technologies with the production systems is further collaborated through the Information and Communication Technologies and Internet of Things. This new framework of key enabling technologies aiding in realization of automation is the main characteristic feature of I 4.0 (Niesen et al. 2016; Schröder et al. 2014).

Integration within I 4.0 can be diversified into horizontal and vertical. Horizontal integration signifies a closer integration between multiple enterprises that have the similar value pertaining to network creation (Brettel et al. 2014). Integration at vertical level indicates collaboration and exchange of information amongst different departments of an enterprise such as management, production and scheduling and planning.

A broader availability and functionality of the various sensor network that are both affordable and efficient is the main condition for successful integration horizontal and vertical levels. Smart and intelligent objects are created that allows for real-time communication between different application systems, working resources and machines. In totality, the entire arrangement taken together, forms the basis for implementation of new automated manufacturing environment and the associated models of business and therefore smart factories (Lucke et al. 2008). The smart manufacturing components are able to acquire and process the relevant data that will aid in self-control of the associated tasks and efficient communication with the human workforce.

Within the milieu of I 4.0 various manufacturing processes are controlled and monitored through the network of sensors and embedded computers. The mechanism of feedback loop affects the physical manufacturing processes as the optimization greatly effects their efficiency. An integrated framework is established between the physical processes and the associated software and networking through the aid of Cyber Physical Systems. The abstraction, modeling and design and analysis of the product manufactured can be carried out in real time. I 4.0 brings about radical changes in the manufacturing environment through changes in the manner various programs are executed. The digital environment allows for real time planning of production schedules and dynamic optimization. Furthermore, the collaboration and cooperation between employees and business partners is made healthier. Looking at the umpteen advantages, nearly 50% of the German firms have adopted the theme of I 4.0 (Sanders et al. 2016).

4.2.1 Literature on Risk Management

Effective management of risk forms a critical component for any enterprise in the present scenario of post-crisis economy. The aspect and knowledge associated with the risk management has been propagated by the Project Management Institute and has been considered as one of the difficult tasks for the project management task and for an enterprise as a whole. In the context of project management, risk management can be defined as a systematic process that aids in identification, analysis and response to the associated risk for streamlining to achieve the desired objective (Banaitiene and Banaitis 2012). Risk management aids the organizations to understand the associated risks, the processes that are under risk, the controlling parameters for the risk and viewing the perspectives of whether the controlling parameters are adequate. If the controlling parameters are not adequate enough, then the actions are required to be undertaken to manage the associated risk and therefore to bring it down to the reasonable and acceptable limits. Efficient risk management framework has become one of the major legal requirements for large organizations which are over and above any moral obligations (Malik and Holt 2013).

Owing to the criticality of the risk management, conceptual framework in the form of Enterprise Risk Management (ERM) has emerged which is developed on the very principles of traditional risk management aspects. ERM has been reported to be a more structured and sophisticated approach that aligns strategy, technology, processes, people that efficiently aids in evaluation and management of the uncertain production environment (KPMG 2001). ISO 31000 encompasses standards that provides guidelines to establish unified and generic conceptual framework for risk management.

Performance measurement forms the basis to the risk management framework. Gaps between the desired level of performance as well as the current performance levels are identified using performance measures. Therefore, performance measures are important as they aid in closing the gaps between both the performance levels. The areas requiring action are suitably identified through Key Performance Indicators and hence the overall performance is enhanced (Weber and Thomas 2005).

Another indicator that aids in detection of risk areas is that of Key Risk Indicator. Investigations have been carried out for the manner in which the Key Risk Indicators aid in detection and hence reduction of the risks (Martin and Power 2007; Scandizzo 2005). Specific risk can be identified through such indicators and hence provide necessary directions in order to monitor the identified risks. These also help in development of warning systems for such risks.

Research lacuna still exists in establishing relationship between various risk indicators and performance indicators. There still exists lack of knowledge that aids in creation of systematic framework for ultimately establishing connection between various indicators. Their effective cooperation can result in providing useful data that can ultimately aid in overall enhancement of company's performance through effective overall risk management.

4.3   Risk Management Framework

The various steps in successful implementation of risk management framework can be divided into: risk identification, design of framework and finally its integration with the performance measures.

4.3.1   Identification of Risk

The major objective of risk identification is the generation of a complete list of potential risks. This is done on the basis of events that might degrade, prevent, enhance, delay or accelerate the process of achieving the desired enterprise' objectives. A comprehensive preparation of risk list is critical as the risk potential that might not be included at this stage will not be analyzed at a later stage. The already existent operational risks associated with manufacturing area are the ones related to maintenance, human sources, material, machining environment, operational methods used, tools and equipment used and manufacturing process used.

The existing potential risks are further accompanied by new emerging risks emanating in the era of I 4.0 through the various threats and the associated vulnerabilities. The major sources of risks are the integrated framework of cyber-space, sophisticated manufacturing technologies, complex elements and the different services of outsourcing. The suitable identification of various risks aid in design and development of risk management framework.

Information security is one of the major reasons for the various operational risks in the manufacturing area. The major challenge lies into protect the manufacturing environment from various cyber-attacks. Prevention of data integrity as well as availability of information are some of the other associated challenges. The challenge can be answered through implementation of information security management systems that is mainly employed by the IT sector.

Information management system is all about maintenance of confidentiality of data i.e., the related information is should only be accessible to the ones who have been authorized. However, the very characteristic feature is just one of the part of the information management system. The other important aspects as already highlighted are that of data availability and its integrity. Availability ensures that the authorized personnel have the access to the information as and when required while on the other hand integrity signifies safeguarding the accuracy of the associated data and hence its completeness. Adoption by various manufacturing enterprises of this system can be a solution to the risks faced by them. The system should however encompass characteristics features of an ISO standard so that it can be a certified system that fulfils the environmental requirements, information and management quality. Furthermore, the information management system must have the potential ability to be effectively integrated into Enterprise Risk Management system.

4.3.2 Design of Risk Management Framework

The next step in successful implementation of the risk management framework is its design. Designed framework must be able to meet the requirements for information management system and that of enterprise risk management system. Once the designed risk management framework has the desired characteristic feature, its safe implementation to the desired enterprise can be ensured. On implementation, the proposed framework could be realized to minimize the risk in the related manufacturing area.

The implementation of the designed risk management framework can well be described through the Deming Plan-Do-Check-Act (PDCA). In the planning phase, the objectives and the vision of an organization are planned. Accordingly, the policies and the procedures are framed to manage the risks associated with the enterprise. Efforts are made to establish improved information security system that can deliver in accordance with the framed policies and objectives of the organization. The next activity in implementation is to Do the processes i.e., to implement and operate the framed policies as well as control the processes and procedures, Subsequently Check is the activity next in the implementation of designed framework. It's quintessential to assess and measure the performance of the processes, objectives and policies set in the Plan phase of the implementation. The results of the assessment should be reported to the management for further review of the obtained results. The last step is to Act i.e., to take steps for improvement and prevention. The management is required to take preventive and corrective actions for the results obtained through internal audit or managerial review. This is necessary to ensure continual enhancement of the performance level of an enterprise.

There should be mechanism in place such that the integrated framework should be well documented, implemented, communicated and improved continuously. Risk management framework should have the potential ability to be transferred into a full-fledged corporate policy. For this it is required to consider the requirements of all the stakeholders involved directly or indirectly. Furthermore, the regulatory as well as legal requirements should be met accordingly.

Effective and functional application of process management must be at the core of implementing the integrated risk management system. Therefore, analysis and optimization of the risk management framework are the keys to effectively monitor the effectiveness of the business process management. Analysis of the associated risk is very important in successful implementation of the proposed risk management framework. The risks identified should be divided into different sections accordingly: technical, planning and processes.

4.3.3 Integration of Performance Measures and Risk Management Framework

Identification of risks and the required actions to be taken to counter the identified risks can be overcome by establishment of suitable business process management plan and the continuity plan. Therefore, the enterprises find it suitable to integrate the business continuity plans and the identified risk treatments into the different processes related to the manufacturing. The measures in place are implemented effectively and tested regularly in order to optimize the performance of the enterprise as a whole. Owing to the criticality of data, risk management has become an integral part of the corporate culture.

The principles for development of framework can be adopted from the fields of Process Performance Management (PPM) and Business Process Management (BPM). The principles are combined with the elements from the framework of risk management. The risk management framework for manufacturing enterprises that is formed on the above basis is developed assuming the following: the identification of risks must be done using clearly defined data and suitable indicators, Important elements of PPM and BPM must be adopted for investigating the performance and level of achievement of objectives and regular monitoring of business processes and their analysis is essential for real-time risk management framework.

Potential threats as well as their occurrence probability can be predicted with greater precision on the basis of data generated from different manufacturing processes. However, to manage the complex scenarios new assessment and examination procedures may be required.

Key risk indicators that influences the key performance indicators aids in monitoring easily different risks and hence the performance of an enterprise. A risk model for the precise analysis of the associate risks might be used. The model can aid in suitably classifying the identified risks into different categories. Different colors can be used for representing the different categorization and prioritization of the identified risk categories. The different categories can be broken down into individual risks.

4.4 Conclusion

The present chapter presents an overview of the aspects of risk management in the context of I 4.0. The aspects related to the implementation of risk management framework have been presented. It can be concluded that the processes of an enterprise can be affected greatly through the human and machine interconnected dynamic and real time self-optimizing systems. However, such complex systems generates great volumes of data and hence their security, integrity as well as real time availability possess major challenge to an enterprise.

Different kinds of risks keep on occurring within an enterprise. The major risk being that of information security. The risks associated are mainly cyber-attacks, loss of integrity of data etc. the frequency of their occurrence is also uncertain. Therefore, the content of the risk management framework will change continuously. Hence existing instruments and infrastructure must be adapted and upgraded as per the requirements.

For suitable application of the risk management framework in light of the enterprise management, a tool connecting the key performance indicators with the key risk indicators must be developed.

References

N. Banaitiene, A. Banaitis, Risk management in construction projects, in *Risk Management-Current Issues and Challenges* (InTech, 2012)

M. Brettel, N. Friederichsen, M. Keller, M. Rosenberg, How virtualization, decentralization and network building change the manufacturing landscape: an industry 4.0 perspective. Int. J. Mech. Ind. Sci. Eng. **8**(1), 37–44 (2014)

M. Hermann, T. Pentek, B. Otto, Design principles for industrie 4.0 scenarios, in *2016 49th Hawaii International Conference on System Sciences (HICSS)* (IEEE, 2016), pp. 3928–3937

L. KPMG, Understanding enterprise risk management: an emerging model for building shareholder value (2001)

M. Lom, O. Pribyl, M. Svitek, Industry 4.0 as a part of smart cities, in *Smart Cities Symposium Prague (SCSP)* (IEEE, 2016), pp. 1–6

S.A. Malik, B. Holt, Factors that affect the adoption of enterprise risk management (ERM). OR Insight **26**(4), 253–269 (2013)

D. Lucke, C. Constantinescu, E. Westkämper, Smart factory—a step towards the next generation of manufacturing, in *Manufacturing Systems and Technologies for the New Frontier* (Springer, London, 2008), pp. 115–118

D. Martin, M. Power, The end of enterprise risk management (Aei-Brookings Joint Center for Regulatory Studies, 2007)

T. Niesen, C. Houy, P. Fettke, P. Loos, Towards an integrative big data analysis framework for data-driven risk management in Industry 4.0, in *2016 49th Hawaii International Conference on System Sciences (HICSS)* (IEEE, 2016), pp. 5065–5074

T.D. Oesterreich, F. Teuteberg, Understanding the implications of digitisation and automation in the context of Industry 4.0: a triangulation approach and elements of a research agenda for the construction industry. Comput. Ind. **83**, 121–139 (2016)

M. Schröder, M. Indorf, W. Kersten, Industry 4.0 and its impact on supply chain risk management, in *14th International Conference "Reliability and Statistics in Transportation and Communi-*

cation (RelStat)" (Riga, 2014), http://www.tsi.lv/sites/default/files/editor/science/Conferences/RelStat14/schroeder_indorf_kersten.pdf

S. Scandizzo, Risk mapping and key risk indicators in operational risk management. Econ. Notes **34**(2), 231–256 (2005)

A. Sanders, C. Elangeswaran, J. Wulfsberg, Industry 4.0 implies lean manufacturing: research activities in industry 4.0 function as enablers for lean manufacturing. J. Ind. Eng. Manag. **9**(3), 811–833 (2016)

A. Weber, R. Thomas, Key performance indicators, in *Measuring and Managing the Maintenance Function* (Ivara Corporation, Burlington, 2005)

Chapter 5
Socio-technical Considerations

5.1 Introduction

Manufacturing environment has been evolving with continuous advancements and research. Digital technologies are being integrated with the manufacturing processes and hence the academicians, practitioners and the researchers considers the current evolving stage as revolutionary phase. Within the manufacturing milieu, Industry 4.0 (I 4.0) has been a topic of debate and research over the recent years. At the core of I 4.0 there are two key enabling technologies: Internet of Things and Cyber-Physical Systems. Learning and understanding of these new technologies aids in realizing an interconnected world wherein customer needs are responded dynamically and efficiently by the developed smart factories. I 4.0 results in the virtual replication of the physical processes and systems which ultimately supports in intelligent analysis of the large volume of generated data. Though up gradation of manufacturing environment through I 4.0 has many associated advantages such as meeting the customer needs through tailoring and customization of products, its adoption by the manufacturing units can't ensure success. The implementation only becomes successful ones it adopts to the architecture or the rules for the prescribed methodology. The designed methodology becomes part of the company culture when a proper relationship establishes between the people receiving the training and organizational capabilities, customers, and stakeholders. Therefore I 4.0 needs to be framed in terms of Socio-Technical Systems in order to reap the desired gains from the implemented model and beyond that by ensuring the system adheres to the technical architectures. In articulating I 4.0 as Socio-Technical System, maximum appreciation from the associated technologies can be realized by the stakeholders engaged directly or indirectly. The relationship between the various entities could also be truly realized. The rules are still being devised for I 4.0 with technical deployment at the center stage. In doing so, the socio-technical aspects are being overlooked. As an important research topic, the present chapter illuminates the readers with the socio-technical characteristics of I 4.0. Chapter is organized to discuss the infrastructure associated with I 4.0 and then goes on to examine as to the possible advantages in adopting I

K. Kumar et al., *Industry 4.0*, Manufacturing and Surface Engineering,
https://doi.org/10.1007/978-981-13-8165-2_5

4.0 environment. The relationship between I 4.0 and lean/six sigma technologies has been assessed next. Chapter illuminates the readers with the socio-technical aspects of I 4.0 before terminating with the concluding remarks.

5.2   Infrastructure for Industry 4.0

Industry 4.0 (I 4.0) has often been considered as phenomenon that results in the paradigm shift in the technology that has been used for the manufacturing processes since the three industrial revolutions. The concept of I 4.0 originated in Germany in the year 2011 and with subsequent advancements and continual research became one of the strategic initiatives of the government. This was included by the German government in the so called "High-Tech Strategy 2020 Action Plan" (Kagermann et al. 2013). This was then promoted as the country's future and was considered as the driving force for the economic and industrial growth (Mosconi 2015).

I 4.0 has been visualized as interconnected framework of digital devices, manufacturing centers, machinery and smart products that can communicate with one another, exchange data and information and hence invoke actions that aids in self-monitoring and controlling of the various elements within the framework. The framework is regarded often as Cyber-Physical System (CPS) (Weyer et al. 2015). Smart factories are made accessible to the heterogeneous network of stakeholders, retailers, customers and suppliers through the high speed internet services. The framework is designed to handle and process volumes of data to ultimately aid in analysis.

I 4.0 has opened a new era of what is called as smart factories (Lee 2015). Smart factories encompasses the machines and systems that interact with the products being manufactured. In the virtual world, these factories are termed as Digital Twin wherein the intelligent self-optimizing algorithms are employed to process the generated data by the different physical elements of manufacturing arena. These self-optimizing algorithms aids in yielding appropriate information with regard to the performance and condition of the various physical systems in real time. Therefore, in the cyber world, the factories have the environment wherein the machines and the other manufacturing equipment are able to analyze themselves their condition through self-diagnosis. This helps the equipment to predict the probability of their malfunctioning and hence their failure. Furthermore, the integrated digital framework helps in dynamic scheduling of the factory in accordance with the customers' requirements and as per the status of the machinery involved in the production.

Internet of Things, Internet of Services and Cyber-Physical Systems are the key components of I 4.0. Collectively, these technologies enables ceaseless communication and effective exchange of information between people, between machines and between machines and people (Roblek 2016). Embedded devices forms the main component of Cyber-Physical Systems. Devices that consists of software and hardware integrated into electrical or mechanical system for performing certain functions have been regarded as embedded devices. Cyber-Physical System is a collection of these devices that not only communicate effectively with one another but also with the physical environment. The communication is realized through the sensors and actu-ators (Alur 2015). Computing facility can be employed more ubiquitously through the aid of Cyber-Physical Systems.

When the medium of communication is internet, this is referred to as Internet of Things. Things are referred to the scenario where the computing capability is extended to the objects and the other related items and products that are often not considered to be digital. The Things exchange and generate data with the minimum human intervention (Xia et al. 2012). Such Things form a network of highly intel-ligent environment that is heterogeneous in nature. The objects within the scenario are not restricted by their geographical location and hence can communicate through internet irrespective of the place where they exist. Internet of Things enable the devel-opment of network comprising of globally distributed suppliers and manufacturing sites.

5.3 Adoption of Industry 4.0

Main question that lies in front of the manufacturers is whether they should embrace the digital environment and hence I 4.0? With the knowledge and understanding on the driving the cost down and enhancing the efficiency, majority of the global manufacturers seems ready to embrace the digital revolution. An average reduction in cost of around 3.5% has been reported by the global survey on I 4.0 held in 2016 by PWC (Global Industry 2016). Added to this reduction cost was also reported the enhanced efficiency of around 4.5% across all the industrial sector. Other obvious benefits of I 4.0 are improved productivity, product quality and enhanced flexibility. Continual improvement and reduced time to market of the manufactured products are also some of the tangible benefits of embracing I 4.0 (Müller 2018). Numbers pertaining to various performance indicators speaks of closeness to that of lean manufacturing methods or six-sigma programs. Therefore, an obvious question of debate amongst academicians and scientific community is that why not adopt and continue with the new evolving technologies and hence the environment.

The answer of the above quest can be deciphered by not only of clear analysis of the benefits but also of the consequences that one needs to face through its non-adoption. The companies that have ignored the adoption of new technologies have either lost their popularity and shares in the market. Subsequently after analyzing the market scenario when they try to bounce back in the market, it becomes difficult for them to again occupy the space of being dominant player. Ultimately such companies then ceases to exist. The observations also apply to the production methods as well as to the technical aspects of the product.

One of the major reason of a firm's failure is the inability of its production system to fulfil the requirements of customer demanding greater variety and attributes in the product (Davies 2014). It has been widely debated that the companies that ignores the adoption of new product technology and focusses only on the present customer needs have overlooked the future requirements of their customer base (Christensen and

Armstrong 1998). Literatures have mentioned the manner in which the manufacturing units maintain their connections with the supplier and customer base. The customer needs have been given due consideration over other business needs by the economists time and again (Levitt 1960). In accordance with the view of Theodore Levitt, the major role for any enterprise to create and maintain the customer base. According to the views of Peter Drucker, one of the well renowned management scientist, enterprises are required to keep themselves align to the market (Drucker 1994) and are required to adapt any future changes in the market. Skinner (Skinner 1969) opines that the manufacturing function should be such that it gives due consideration to customer needs over the minimization of cost. In the latter half of the 20th century, when there was decline in the American manufacturing, the research community continued to propagate and promote manufacturing as a competitive lever and proposed many models such as four stage model by Hayes and Wheelwright (Wheelwright and Hayes 1985). Different manufacturing models were proposed with the sole objective to reverse the declining manufacturing trend that was prominent during 70s and 80s (Hayes and Wheelwright 1984).

With the passage of time, the proposed four stage model received wider acceptance and thereby gaining popularity in the domain of operations management (Barnes and Rowbotham 2004). It is believed that the consequences of non-adoption of I 4.0 environment can be perceived on reversing the directional movement in the four stage manufacturing model. The current dominant player in the market will not be able to meet the expectations of the redefining industries and therefore will not remain best in their field in the near future. Non-implementation of I 4.0 by an enterprise will ultimately halt the impact of their manufacturing operations strategically and ultimately their organization will remain behind in the competitive market.

5.4 Lean Methodologies and Industry 4.0

Examinations and analyses have been carried out to establish relationship between I 4.0 and lean manufacturing methodologies (Sanders et al. 2016). Lean manufacturing is methodology that focusses on the customer base to deliver them with the desired product value and attributes with the employability of minimum resources. Six Sigma on the other hand is a problem solving methodology that is based on the following: Define, Measure, Analyse, Improve and Control (DMAIC). Both methodologies are being employed in tandem by various companies and organizations to effectively deliver improvements for their business (Maleyeff et al. 2012).

There are number of benefit associated with joint implementation of Six Sigma and Lean manufacturing methods other than delivering desired customer value within cost constraints. These benefits are however less tangible but can be perceived as key enablers for I 4.0 implementation. Creation of an environmental culture that motivates for continual improvement aids the managerial staff and the other related workforce to not only accept and adapt to the changes but also actively drive the changes. An embedded problem solving structure will also help the analysis staff to implement sound, scientific and sustainable solutions to any arising problem. Furthermore, the production systems will be efficient, stable and productive owing to the minimal rejections, defects and delays.

Thus, implementation of both the Lean and Six Sigma methodologies will enable an environment for implementation of I 4.0 and therefore a step in leveraging enhanced operational and enterprise performance. On the other hand, I 4.0 provides the potential infrastructure to Lean and Six sigma methodologies so that organizational performance is enhanced effectively.

The data and metrics related to the organizational performance is transmitted effectively in real time with the employability of Cyber-Physical Systems. This could include the following: Data Analysis, Production Surveillance, Virtual Value Stream Mapping, Total Productive Maintenance and Electronic Kanbans. The data generated in real time by various manufacturing processes and machineries are stored for analysis at later stage. Some statistical methods may be employed to the stored data which ultimately will aid in continuous efficiency improvement of various related activities. Behavior of the manufacturing systems can also be analyzed over different times through the utilization of different predictive analytical tools. Through Production Surveillance, key production metrics such as set up time, downtime and production rates are captured automatically. Production related problems such as deterioration in the performance of machines and failure of machines are anticipated in advance through intelligent protocols. The related information are communicated effectively to the personnel involved in the production through the employability of smart devices such as tablets and smart watches. Value stream mapping is another core activity encompassed within the lean initiative framework (Rother and Shook 2003). Virtual value stream mapping can be established with the aid of Virtual Reality (Fuchs 2017). The employability of Virtual Reality environment eliminates the need to understand Virtual value stream mapping framework. Different stakeholders can be involved in the virtual value stream that can do away with the understanding of interaction amongst the virtual value stream mapping symbols. The stakeholders could be able to observe clearly the current and future models of the manufacturing state. Total Productive Maintenance encompasses sensors that detect when a particular component needs to be replaced or when replenishment of oil is desired. Accordingly signals are sent to the concerned personnel for necessary action. An alert signal can also be generated in advance being triggered by the calendar date. The framework could also include electronic Kanban cards instead of the conventional ones. These can aid in automatically transferring the production order to the downstream process on sensing the inventory level.

Lean enterprises is considered to be a suitable route to lean thinking that leads to the creation of extended supply by connecting manufacturers, retailers and the customers (Womack and Jones 1997). Lean enterprises have been considered to be the key drivers by Lean Aerospace Initiative in reducing the cost of procurement and improving the time of delivery for the aerospace sector (Murman et al. 2002). Lean Aerospace Initiative is a connected network of aerospace suppliers, U.S. Air force, commercial airlines, aircraft manufacturers and academicians. The variety, volume and the speed of information and data exchange is constrained by the technologies available.

The connecting capability of I 4.0 framework can promote the exchange and transfer of data with variety, volume and speed. The only restricting concern can be that of protocols related to security and authorization. The connecting ability helps in realizing the vision of lean enterprise.

5.5 Socio-technical Systems for I 4.0

Reference Architectural Model for I 4.0 is a technical architecture for I 4.0 that has been proposed by Platform Industrie 4.0 (Pîrvu and Zamfirescu 2017). The proposed model can be visualized as three dimensional coordinate system consisting of different layers in series. Each layer is representative of different manufacturing functions. A business layer is also one of the layers within the model, but more value can be added by consideration of the socio-technical features of I 4.0.

Socio-technical systems are the ones that encompasses relationship between machines, humans and the aspects related to the environment (Baxter and Sommerville 2011). Social and technical interactions have been considered as an important aspect of practical relevance that supports and promotes for organizational developments and adoption of changes (Davis et al. 2014).

The different paradigm industrial shifts have evidenced the new working ways and were prominently involved in creation of new job roles and manufacturing disciplines. Factory system of manufacturing was created in the aftermath of classic industrial revolution. Mass production ushered in the era of division of workforce and elevated the conceptual framework of scientific management. Disciplines such as Industrial Engineering and Operations Management was born as a consequence of mass production era. Industrial revolution with primary focus on automation resulted in transferring the responsibilities of manual workers to that of control worker. The transformation gave rise to another discipline referred to as control engineering.

In the present scenario, I 4.0 is still evolving through the current knowledge base. Still research is under progress regarding the impact of I 4.0 on the workforce, the interaction of people involved with the technologies and on the new ways of working and emerging disciplines (Davies et al. 2017). Substantial changes can be perceived through the employability of digital technologies that support and promote I 4.0 environment. The socio-techno relationship can be conceived to have more degrees of freedom in comparison to the conventional technologies owing to the heterogeneous characteristics of the integrated network (Kopp et al. 2016).

Socio-technical interaction results in increasing complexities that can be managed by the way in which the personnel at each organization level interact with the socio-technical system in place. The personnel belonging to the executive level of an organization will have to have a better relationship with the personnel at operational level. This is because the executive personnel will have to rely heavily on the lower tier personnel to accurately understand the operational status. Executive personnel are required to know the questions that needs to be asked and also the answers to be given to the questions asked to them. Maintenance of healthy relationship is necessary so that effective connectedness is maintained with the heterogeneous customer base. The executive level must be able to make strategic decisions as and when the trends and requirements of the market and the customer base changes. There should be active engagements between the management and the workforce they control. The engagements should allow for effective transfer of knowledge between the management and the different levels of the organization. Shared knowledge will aid in optimizing the decisions taken by the management. The sharing of knowledge should be done so as to conceive the hierarchy divisions as blurred. The personnel at the operational levels will not be passive rather their status will be elevated to that of knowledge worker.

As the future will unfold, there will be a more clear understanding of the emerging disciplines with the adoption of I 4.0 environment. The various personnel involved include technical experts, managers, knowledge workers and data analysts. Further requirement is demanded for to have a knowledge professional who will aid in merging key elements of the various disciplines and hence is able to provide a better viewpoint of the technical system as a whole. The job role of such professionals will be multi-faceted that will entail providing advice to the various organizational disciplines on the course of action to be taken by them so that a precise alignment is maintained within the heterogeneous and integrated framework.

5.6   Conclusion

The policies that considers driving down the costs and increasing the organizational efficiencies should always be given due consideration. I 4.0 is one such evolving technology that matches on these metrics. However, the efficient connectedness with the dynamic and heterogeneous customer base is the key to the successful implementation and adoption of I 4.0. This can be achieved through the employability of appropriate analytical methods. The present chapter explores the integrated framework of I 4.0 encompassing lean and six sigma methodologies. The methodologies have been revealed to be mutually supportive. However, rigorous research is still required in this domain to establish a well proven business model.

Further, successful implementation of technical architectural structure of I 4.0 will also rely on the socio-technical interactions. Research and investigations needs to be done in this direction to fully comprehend the impact of socio-technical interactions. It is clear that the manner in which the all the stakeholders work needs to be changed. The conventional management will have to be transformed to knowledge source. The role of knowledge professional in streamlining the knowledge transfer process has been appropriately highlighted in the present context.

References

R. Alur, *Principles of Cyber-Physical Systems* (MIT Press, 2015)

D. Barnes, F. Rowbotham, Testing the four-stage model of the strategic role of operations in a UK context. Int. J. Oper. Prod. Manag. **24**(7), 701–720 (2004)

G. Baxter, I. Sommerville, Socio-technical systems: from design methods to systems engineering. Interact. Comput. **23**(1), 4–17 (2011)

C.M. Christensen, E.G. Armstrong, Disruptive technologies: a credible threat to leading programs in continuing medical education? J. Contin. Educ. Health Prof. **18**(2), 69–80 (1998)

R.S. Davies, A structured approach to modelling lean batch production (Doctoral dissertation, Buckinghamshire New University, 2014)

P.F. Drucker, *The Theory of Business*. (Harvard Business Review, Boston, 1994), p. 95

M.C. Davis, R. Challenger, D.N. Jayewardene, C.W. Clegg, Advancing socio-technical systems thinking: a call for bravery. Appl. Ergon. **45**(2), 171–180 (2014)

R. Davies, T. Coole, A. Smith, Review of socio-technical considerations to ensure successful implementation of Industry 4.0. Procedia Manuf. **11**, 1288–1295 (2017)

P. Fuchs, Virtual Reality Headsets-A Theoretical and Pragmatic Approach (CRC Press, 2017)

Global Industry 2016. 4.0 Survey—Industry Key Findings. Accessed 10 Sept 2017, www.pwc.com/industry40

R.H. Hayes, S.C. Wheelwright, *Restoring Our Competitive Edge: Competing Through Manufacturing*, vol. 8 (Wiley, New York, NY, 1984)

H. Kagermann, J. Helbig, A. Hellinger, W. Wahlster, Recommendations for implementing the strategic initiative Industrie 4.0: securing the future of German manufacturing industry; Final report of the Industrie 4.0 Working Group. Forschungsunion (2013)

R. Kopp, J. Howaldt, J. Schultze Why Industry 4.0 needs workplace innovation: a critical look at the German debate on advanced manufacturing. Eur. J. Workplace Innov. 2(1) (2016)

T. Levitt, Marketing myopia (1960), pp. 45–56

J. Lee, Smart factory systems. Informatik-Spektrum **38**(3), 230–235 (2015)

J. Maleyeff, E.A. Arnheiter, V. Venkateswaran, The continuing evolution of lean six sigma. TQM J. **24**(6), 542–555 (2012)

F. Mosconi, The new European industrial policy: global competitiveness and the manufacturing renaissance (Routledge, 2015)

R. Müller, M. Vette-Steinkamp, L. Hörauf, C. Speicher, D. Burkhard, Development of an intelligent material shuttle to digitize and connect production areas with the production process planning department. Procedia CIRP **72**, 967–972 (2018)

E. Murman, T. Allen, K. Bozdogan, J. Cutcher-Gershenfeld, H. McManus, D. Nightingale, E. Rebentisch, T. Shields, F. Stahl, M. Walton, J. Warmkessel, Lean enterprise value. *Insights From Mit's Lean* (2002)

B.C. Pîrvu, C.B. Zamfirescu, Smart factory in the context of 4th industrial revolution: challenges and opportunities for Romania, in *IOP Conference Series: Materials Science and Engineering*, vol. 227, no. 1, p. 012094 (IOP Publishing, 2017)

V. Roblek, M. Meško, A. Krapež, A complex view of industry 4.0. Sage Open **6**(2), 2158244016653987 (2016)

M. Rother, J. Shook, Learning to see: value stream mapping to add value and eliminate muda (Lean Enterprise Institute, 2003)

A. Sanders, C. Elangeswaran, J. Wulfsberg, Industry 4.0 implies lean manufacturing: research activities in Industry 4.0 function as enablers for lean manufacturing. J. Ind. Eng. Manag. **9**(3), 811–833 (2016)

W. Skinner, Manufacturing-missing link in corporate strategy (1969)

S.C. Wheelwright, R.H. Hayes, Competing through manufacturing. Harvard Bus. Rev. **63**(1), 99–109 (1985)

S. Weyer, M. Schmitt, M. Ohmer, D. Gorecky, Towards Industry 4.0-standardization as the crucial challenge for highly modular, multi-vendor production systems. IFAC-PapersOnline, **48**(3), 579–584 (2015)

J.P. Womack, D.T. Jones, Lean thinking—banish waste and create wealth in your corporation. J. Oper. Res. Soc. **48**(11), 1148–1148 (1997)

F. Xia, L.T. Yang, L. Wang, A. Vinel, Internet of things. Int. J. Commun. Syst. **25**(9), 1101–1102 (2012)

Chapter 6
Sustainable Business Scenarios in 4.0 Era

6.1 Introduction

Customer needs and business are not only the sole sources of challenges for any business model. There has been unprecedented growth of business post-world-war era, however, there has been a greater degree of misbalance between the demand and supply. This is because of the purchasing power of each individual has become threefold (Mundial 2015; Stock and Seliger 2016). The increasing purchasing demand has resulted in tremendous pressure on the resources and the global climate has been affected greatly ultimately leading to social instability (Hsiang and Burke 2014). Studies on environmental sustainability exist post the groundbreaking research by club of Rome in the 70s (Turner 2012).

Scientific community opines that classical business models cannot be underpinned by sustainability. These classical business models are required to be modified to include factors that involve social and environmental considerations (Stubbs and Cocklin 2008). The old design practices have resulted in customers going for the product replacement without giving due consideration to repair, recycle or dispose the product for which the replacement is sought (Guiltinan 2009). Relationship between business models and sustainable innovation has revealed four major missing elements in the context of sustainable innovation: financial model, customer interface and supply chain and value proposition. Although adoption of integrated sustainable business models is profitable, however, for industries that are concentrated on only being financial successful ignores the very outlook of being sustainable (Boons and Lüdeke-Freund 2013).

Being sustainable is one of the major challenges that is faced by the industrial units. Apart from being sustainable, adoption of key technological advancements in the form of automation and digitization are the other associated problems. The era of digitalization and automation is referred to as Industry 4.0 (I 4.0) (Lasi et al. 2014). I 4.0 is a conceptual framework that encompasses integration amongst various horizontal levels of any organization. The different elements of this horizontal integration include products, customers, suppliers, workers and manufacturing equipment. All

K. Kumar et al., *Industry 4.0*, Manufacturing and Surface Engineering, https://doi.org/10.1007/978-981-13-8165-2_6

these discrete elements are embedded within a virtual network. Exchange of data takes place between the various elements of the virtual network through embedded devices (Seliger et al. 2016). Another outlook towards I 4.0 is that it is network of different manufacturing resources that are self-monitoring, self-optimizing and self-controlling. The manufacturing resources are knowledge-based and equipped with sensors that aid these resources to adopt to the aforementioned characteristic features (Kagermann et al. 2013).

The present chapter aims to address the influence of I 4.0 on the business models that promote sustainability. Modelling of business models incorporating sustainability and I 4.0 has been discussed. Various business scenarios on the integrated model have also been highlighted. The chapter illuminates on the future research directions before terminating with the concluding remarks.

6.2   Sustainable and Digital Business Models

The following section discusses the application of sustainable innovations and I 4.0 to the existing business models. The reflection of sustainability and I 4.0 has hen been discussed after the discussion on above has been done.

6.2.1   Sustainability and Supply Chain

A supply chain is always required to deliver the value of proposed business model. A supply chain transforms the input to a higher value output. The present section discusses on sustainable supply chain proposed by de Man and Strandhagen (2017). They considered a process model given by Holweg (Boer et al. 2015). A feedback loop was considered for repair, reverse logistics, reclaim, refurbishing and recycling (Srivastava 2007). To lend sustainability characteristics to the model, the three main production strategies consistency, sufficiency and efficiency are linked to the model (de Man and Friege 2016). Further, the process model is connected with the natural environment providing for an integrated supply chain with environmental considerations. This has been achieved by connecting the process model with the raw material extraction process, waste creation framework and to the input of energy, water and other regenerative processes so that the natural environment is restored. Sufficiency is obtained by reduction of the inputs, efficiency by making use of suitable raw material extraction process and their efficient transformation. Consistency of the model can be ensured with the feedback loop being able to recycle and bring back the material again to the supply chain.

Various strategies related to the valuation of environmental impact are used to measure the effect of sustainability strategies. Some of these include life cycle inventory analysis, life cycle impact analysis and life cycle assessment (Turner 2012). Therefore, through the above discussion, it is clear that supply chains can be integrated with sustainability strategies and the impact can be assessed by measuring environmental impacts that the products will have.

6.2.2 Industry 4.0 and Its Applications

Opportunities have been identified by the scientific community for the implementation of sustainable manufacturing models (Seliger et al. 2016). Means to exploit data to be used for the sustainable business models and the framework for product life cycle with feedback loop are some of the identified opportunities. However, the identified opportunities did not translate into tangible benefits and hence the future research agenda. Although it is understood that I 4.0 can help establish sustainable business models, it is still unclear as to how this can be achieved.

The data can be collected from a wide range of sources ranging from customer to the raw material. However, until and unless the collected data is exploited, the proposition of value creation cannot be realized truly.

Lacuna still exists to translate the business model devoted to sustainable strategy. Gap continues to exist to establish a mechanism that can aid in using the collected data from the product's life cycle for obtaining sustainable products. Although implementation of I 4.0 environment results in better service involving higher quality products, mass customization, faster and efficient production systems but I 4.0 has failed to tackle explicitly the sustainability issues faced by the manufacturers. Sustainability is often tackled implicitly when the benefits from sustainable innovation reaps economic benefits. As for example, if for a particular company its turnover from the sale of the product remains similar but it increases from the services then it does not mean explicitly that the company has become more sustainable.

6.2.3 Reflection of Sustainability on Business Models

There is a debating question amongst the scientific and management community on how to run efficiently design a sustainable business model and hence running it efficiently. Although there are multiple examples of sustainable business models (de Man and Strandhagen 2017) in the marketing space, but they only have a very small market share and they cater to the niche customer base. Value proposition has been turned to be one of the major reasons for customers moving from one company to another (Osterwalder and Pigneur 2010). The needs of the customer and their problems are satisfied through value proposition. Durability of the product is often not conceived as an environmental concern by majority of the customer base and there will be no drastic increase in the consumer base that supports green products (Cooper 2004; Hellweg and i Canals 2014). Hence it is very difficult to create value proposition with due consideration to sustainability that will ensure a larger market base.

A decrease in product life cycle has been reported for the consumer goods even though there is promising idea from the proponents of sustainable business models to offer products with enhanced life through better services. Innovation is one of the primary reasons behind decrease in the product deployment time (Guiltinan 2009).

There are practical and theoretical limitations in adopting circular economics that promotes for recycling and reuse (de Man and Friege 2016). Conceptual framework of circular economics if made feasible can result in production of newer products without the need of newer materials. Hence most of the models of business do not support sustainable supply chain operation that simultaneously can meet the customer demand and is justifiable financially.

The viewpoint of sustainable supply chain is the reduction of newer raw materials by promoting recoverability through durable products that are designed for reusability, recyclability and refurbish ability. The implementation of I 4.0 environment will aid in deciphering the limits of the parameters within which the products and the manufacturing processes can function properly. Further, the signs of wear and tear are easily recognizable (Kagermann et al. 2013) so that necessary preventive action could be taken.

Challenges are posed by the products that have longer lifetime in the context of profitable models of business. One such challenge includes the management of relationship between product end-of-life, customer and the company. This may arise particularly when companies focus for short-term and have the probability to go bankrupt. Another important challenge is that whether value proposition offered by sustainable business model is competitive enough to that of the classical business model.

6.2.4 Integration of I 4.0 with Sustainability

The integration of I 4.0 with the sustainable business model results in physical and digital connectedness of the customer base with the product. The supply chain and customer is also connected digitally through I 4.0. Value proposition remains to be central for such integrated business model that balances the economic, social needs and ecological needs. These values are determined spatially and temporally (Boons and Lüdeke-Freund 2013). However, financial justification still needs to be looked at for such digital and sustainable business models.

6.3 Scenarios Countering Strategies for Obsolescence

The two potential scenarios that counters for obsolescence have been discussed in this section. These obsolescence are related to design for improvement in product functionality by addition or up gradation of the existing product, designing the product that promotes for limited repairs and designing the product for aesthetics that results in reduced customer satisfaction. The two scenarios are related to kitchen appliances and laptop business.

6.3.1 Kitchen Appliances

Various equipment in the kitchen requires to be renovated and renewed depending on the individual purchase and repair of each product. These are based on the requirement of maintenance, requirement of repairs and inspections and repairs. Limited functional life of the product, designed aesthetics that results in reduced customer satisfaction and designing of product that has limited repairs are the major reasons behind the product becoming obsolete.

To counter the obsolescence, one could propose value proposition that can offer modular products. These products would be surveyed through sensors, integrated with Internet of Things allowing for remote supervision ultimately leading to efficient usage, longer lifetime and planned maintenance.

The current practices of maintenance can be replaced with smarter maintenance protocols and repairing solutions. The items that have upgradable and interchangeable units allow for renovation with least disposal of materials and removal of any unit. The smart and modular products allow the customers to track the status of their products through energy efficiency measurements.

Long-term commitments are demanded from these modular products, and hence business solutions. Wider range of products are also required to truly realize the solutions. Furthermore a business space is required wherein such products can be leased out.

The business solution can be justified financially when service contracting is made feasible wherein the customer pays for the maintenance and monitoring of the product. Contracting services may also be adopted for recycling the products that fails during their service time.

A product will contribute towards sufficiency when it arises as a result of more long term relationship. The product could also meet the efficiency requirements if it monitored effectively.

6.3.2 Laptop Business

In general, the current lifespan of a laptop ranges 3–6 years. Planned obsolescence can be considered as one of the major reason behind the shorter lifespan. Destabilization of the specifications and qualities in the laptop results in customers constantly requalifying for different version of the products. It is easier to notice the technological changes over the years, but there has not been much changes to the main features and the usage. Currently, because of the non-modular setup, the laptops are difficult to repair. Majority of the software are in cloud and therefore the laptops should be provided with screen and keyboards connected to the Internet and processor.

A value proposition can be to provide laptops with internet connections and access to cloud. This will aid in eliminating the need to replace the entire laptop again and again. Such laptops already exist in the form of thin clients. However, these smart laptops may not promote sustainability wherein these could be recycled or remanufactured. Leasing out of the thin clients can be a very good instance to promote sustainability. This can be achieved as this entails the usage of lesser raw materials and thereby promoting sufficiency. A more efficient supply chain is achieved and also consistency wherein the customer is encouraged to return the product and hence recycling.

6.4   Future Research Directions

Delivering of sustainable value proposition opens the frontiers of research. This may encompass to incorporate various elements of I 4.0 into theoretical foundation and hence promoting sustainable value proposition. Research needs to be carried out to analyze the changes that will occur because of the competition between sustainable and non-sustainable products. Management of supply chain is another future research direction. This very domain may encompass to carry out investigations on the impact that the smart factories and products will have on the network of supply chain. Reverse logistics based supply chain is another frontier open to research in this very field. Loyalty of customer to motivate sustainable business practices is another research frontier. Cost–benefit analysis should be carried out to judge the sustainability of the supply chain and hence the business model as a whole.

6.5   Conclusion

The present chapter illuminates the readers with the concept of sustainable business models in the era I 4.0. Conceptual framework of sustainable and digital business model was elaborated. Some of the potential business scenarios were highlighted. Furthermore, the challenges associated with the business models vouching for design for obsolescence have been outlined. Although the challenges exist, the potentiality of sustainable business models in era I 4.0 exists. A complete market shift towards sustainability will altogether depend on the support that would be provided by the key enabling I 4.0 technologies.

References

H. Boer, M. Holweg, M. Kilduff, M. Pagell, R. Schmenner, C. Voss, Making a meaningful contribution to theory. Int. J. Oper. Prod. Manag. **35**(9), 1231–1252 (2015)

F. Boons, F. Lüdeke-Freund, Business models for sustainable innovation: state-of-the-art and steps towards a research agenda. J. Clean. Prod. **45**, 9–19 (2013)

T. Cooper, Inadequate life? Evidence of consumer attitudes to product obsolescence. J. Consum. Policy **27**(4), 421–449 (2004)

R. de Man, H. Friege, Circular economy: European policy on shaky ground (2016)

J.C. de Man, J.O. Strandhagen, An Industry 4.0 research agenda for sustainable business models. Procedia CIRP **63**, 721–726 (2017)

S. Hellweg, L.M. i Canals, Emerging approaches, challenges and opportunities in life cycle assessment. Science **344**(6188), 1109–1113 (2014)

S.M. Hsiang, M. Burke, Climate, conflict, and social stability: what does the evidence say? Clim. Change **123**(1), 39–55 (2014)

J. Guiltinan, Creative destruction and destructive creations: environmental ethics and planned obsolescence. J. Bus. Ethics **89**(1), 19–28 (2009)

H. Kagermann, J. Helbig, A. Hellinger, W. Wahlster, Recommendations for implementing the strategic initiative INDUSTRIE 4.0: securing the future of German manufacturing industry; final report of the Industrie 4.0 Working Group. Forschungsunion (2013)

H. Lasi, P. Fettke, H.G. Kemper, T. Feld, M. Hoffmann, Industry 4.0. Bus. Inf. Syst. Eng. **6**(4), 239–242 (2014)

B. Mundial, *World Bank Open Data*. The World Bank Group (2015), http://data.worldbank.org

A. Osterwalder, Y. Pigneur, *Business Model Generation: A Handbook for Visionaries, Game Changers, and Challengers* (Wiley, 2010)

R. Seliger, G. Hartz, E. Fontana, D. Fusari, U.S. Patent No. 9,392,078. Washington, DC: U.S. Patent and Trademark Office (2016)

S.K. Srivastava, Green supply-chain management: a state-of-the-art literature review. Int. J. Manag. Rev. **9**(1), 53–80 (2007)

T. Stock, G. Seliger, Opportunities of sustainable manufacturing in industry 4.0. Procedia Cirp **40**, 536–541 (2016)

W. Stubbs, C. Cocklin, Conceptualizing a "sustainability business model". Organ. Environ. **21**(2), 103–127 (2008)

G.M. Turner, On the cusp of global collapse? Updated comparison of the limits to growth with historical data. GAIA-Ecol. Perspect. Sci. Soc. **21**(2), 116–124 (2012)